Springer Theses

Recognizing Outstanding Ph.D. Research

For further volumes:
http://www.springer.com/series/8790

Aims and Scope

The series "Springer Theses" brings together a selection of the very best Ph.D. theses from around the world and across the physical sciences. Nominated and endorsed by two recognized specialists, each published volume has been selected for its scientific excellence and the high impact of its contents for the pertinent field of research. For greater accessibility to non-specialists, the published versions include an extended introduction, as well as a foreword by the student's supervisor explaining the special relevance of the work for the field. As a whole, the series will provide a valuable resource both for newcomers to the research fields described, and for other scientists seeking detailed background information on special questions. Finally, it provides an accredited documentation of the valuable contributions made by today's younger generation of scientists.

Theses are accepted into the series by invited nomination only and must fulfill all of the following criteria

- They must be written in good English.
- The topic of should fall within the confines of Chemistry, Physics and related interdisciplinary fields such as Materials, Nanoscience, Chemical Engineering, Complex Systems and Biophysics.
- The work reported in the thesis must represent a significant scientific advance.
- If the thesis includes previously published material, permission to reproduce this must be gained from the respective copyright holder.
- They must have been examined and passed during the 12 months prior to nomination.
- Each thesis should include a foreword by the supervisor outlining the significance of its content.
- The theses should have a clearly defined structure including an introduction accessible to scientists not expert in that particular field.

Dag Gillberg

Discovery of Single Top Quark Production

Doctoral Thesis accepted by
Simon Fraser University,
Burnaby, BC, Canada

 Springer

Author
Dr. Dag Gillberg
Carleton University
Arthur Street 1–141
Ottawa ON K1R 7C1
Canada
e-mail: dag.gillberg@cern.ch

Supervisor
Dugan O'Neil
Department of Physics
Simon Fraser University
888 University Drive
Burnaby BC
Canada

ISSN 2190-5053

e-ISSN 2190-5061

ISBN 978-1-4614-2798-8

ISBN 978-1-4419-7799-1 (eBook)

DOI 10.1007/978-1-4419-7799-1

Springer New York Dordrecht Heidelberg London

Cover design: eStudio Calamar, Berlin/Figueres

Printed on acid-free paper

Springer is part of Springer Science+Business Media (www.springer.com)

Supervisor's Foreword

This thesis documents one of the flagship measurements of the D0 experiment, a collaboration of more than 600 physicists from around the world. It describes first observation of a physical process known as "single top quark production", which had been sought for more than 10 years before its eventual discovery in 2009. Further, the thesis describes, in detail, the innovative approach Dr. Gillberg took to this analysis. Through the use of Boosted Decision Trees, a machine-learning technique, he observed the tiny single top signal within an otherwise overwhelming background. This document should serve not only as the definitive guide to an important experimental result, but also as a valuable text with regards to the application of Boosted Decision Trees to problems in particle physics.

Unlike top quark pair production, single top quark production occurs via the electroweak interaction. One consequence of this is that the the process allows us to make unique measurements of top quark properties. In particular, the first direct measurement of the Cabibbo-Kobayashi-Maskawa mixing matrix parameter $|V_{tb}|$ was derived in the context of this work. The general interest in the work is evident in the high number of citations of the related papers on single top quark evidence and observation. In addition, each paper was made an "editor's choice" in Physical Review Letters and the observation was highlighted in the journal Physics as one of the most interesting and significant results published in all of physics at that time. This thesis does a far more detailed and complete job of documenting the techniques and results used in these measurements than any existing published paper.

The trouble with single top quarks is that they are masked by two distinct and difficult backgrounds, each with different characteristics. We often say that single top quark signatures are kinematically "wedged" between top quark pair production and W boson production backgrounds. As such, the search for single top quarks proved to be exceedingly difficult. There were two critical ingredients to making the measurement: (1) a detailed model of the D0 detector and background processes and (2) an advanced multivariate approach to extracting the signal from the backgrounds. This thesis illustrates both of these important aspects. Dr. Gillberg was a key player both in understanding the D0 dataset and in developing innovative techniques to "mine" the information from that dataset. This is evident in his written work.

This thesis serves as the definitive documentation of the observation of single top quark production. It also already being used as a reference for people learning the application of the Boosted Decision Tree technique to particle physics. The impact of this work goes beyond the initial measurement, it includes a set of technical details with widespread applicability in the field. I am very pleased to see it published in Springer Thesis series.

Burnaby, July 2010   Dugan O'Neil

Preface

The top quark is by far the heaviest known fundamental particle with a mass nearing that of a gold atom. Because of this strikingly high mass, the top quark has several unique properties and might play an important role in electroweak symmetry breaking—the mechanism that gives all elementary particles mass. Creating top quarks requires access to very high energy collisions, and at present only the Tevatron collider at Fermilab is capable of reaching these energies.

Until now, top quarks have only been observed produced in pairs via the strong interaction. At hadron colliders, it should also be possible to produce single top quarks via the electroweak interaction. Studies of single top quark production provide opportunities to measure the top quark spin, how top quarks mix with other quarks, and to look for new physics beyond the standard model. Because of these interesting properties, scientists have been looking for single top quarks for more than 15 years.

This thesis presents the first discovery of single top quark production. An analysis is performed using 2.3 fb^{-1} of data recorded by the DØ detector at the Fermilab Tevatron Collider at centre-of-mass energy $\sqrt{s} = 1.96$ TeV. Boosted decision trees are used to isolate the single top signal from background, and the single top cross section is measured to be $\sigma\,(\bar{p}p \rightarrow tb + X,\, tqb + X) = 3.74^{+0.95}_{-0.74}$ pb.

Using the same analysis, a measurement of the amplitude of the CKM matrix element V_{tb}, governing how top and b quarks mix, is also performed. The measurement yields: $|V_{tb}f_1^L| = 1.05^{+0.13}_{-0.12}$, where f_1^L is the left-handed Wtb coupling. The separation of signal from background is improved by combining the boosted decision trees with two other multivariate techniques. A new cross section measurement is performed, and the significance for the excess over the predicted background exceeds 5 standard deviations.

Acknowledgments

To Dugan O'Neil for guidance and support through both a Masters' and a Ph.D., always ensuring the focus was maintained on what was most needed. Dugan, thank you for showing me how an adviser should be.

To Yann Coadou for schooling in the arts of careful and thorough high energy physics analysis.

To Marco Bieri for convincing me to start with experimental particle physics, the Cambie pub nights and the music.

To Mike Vetterli and Reda Tafirout for their advice throughout this project, and for not ripping me apart during the defence

To the single top conveners: Reinhard for being the resident smart guy; Ann for always noticing and caring about the details; Cecilia for her widespread knowledge; and Aran for making working with single top a great experience.

To the group of grad students and postdocs at DØ. In particular to my BDT partner Jorge, and to Monica, Andres, Gustavo and Liang—without you guys the single top analysis would still be at square one. Also to Jiri, Krisztian and the Fermilab football crew for the occasional time off work while at the lab.

To my fellow grad students and postdocs at SFU, especially: Zhiyi for his humour and our lively discussions; Doug for insightful conversations; Jen for allowing herself to be the object of endless teasing; Erfy-Derfy for the football games in the hallway; and, Travis for our common love of beer.

To friends and family in Sweden: the "iron gang" members Mikael, Henrik and Andreas for always being there despite my dismal communication; to mamma and pappa for raising me to realize that life is an adventure about to be explored; and, to my siblings, Ann, Linn, Tor and Love, for all the shared adventures, whether perceived or real.

Finally to Michelle, for her support, encouragement and always being there, for doing particle physics, and feeding me with the most delicious food. Mmm... I love you.

Contents

Chapter 1
Introduction

To the best of our understanding, all observed physical phenomena can be explained by four fundamental interactions (forces). Our current theory of elementary particle physics, the *standard model*,[1] incorporates three of the four forces and can accurately describe all experimental observations to date. Even though this model has been remarkably successful, it is widely believed that it is a low energy approximation of a more profound theory. Hints of this theory are expected to be observed at the very highest energies.

In order to probe the fundamental nature of our world, one needs to build very large and powerful machines. At the time of this analysis, the world's highest energy particle accelerator was the Tevatron proton–antiproton collider, located at Fermilab outside Chicago. This title is currently held by the Large Hadron Collider (LHC), near Geneva, Switzerland. These are the only two machines powerful enough to produce *top quarks*.

The top quark is by far the heaviest fundamental particle in the standard model. Because of its large mass, the top quark has several unique properties and could provide hints for the origin of mass and physics beyond the standard model. At the Tevatron and LHC, top quarks are predicted to be produced in pairs via the strong force, and singly via the electroweak force. Top pair production was discovered by the DØ and CDF collaborations in 1995. This thesis presents the first observation of electroweak single top quark production.

When single top quark production occurs, it leaves a characteristic "footprint" in the DØ detector. Unfortunately, several background processes leave a very large amount of similar footprints, and one can never be absolutely sure whether the footprint left by a particular collision originate from single top quark production or not. In order to prove its existence, one must identify the most distinguishing features between single top quark footprints and background footprints. This analysis successfully uses boosted decision trees for this purpose. After the

---

[1] For basic introductions of the standard model and particle physics in general, see Refs. [1] and [2].

D. Gillberg, *Discovery of Single Top Quark Production*, Springer Theses,
DOI: 10.1007/978-1-4419-7799-1_1, © Springer Science+Business Media, LLC 2011

decision trees are applied to the collision data, a statistical analysis is performed to measure the rate of single top quark production, and to establish its existence.

The outline of this thesis is as follows. Chapter 2 provides a theoretical introduction and motivation for the study of single top quark production. The following chapter describes the experimental apparatus, namely the Fermilab accelerator chain and the DØ detector. Chapter 4 explains how signals in the DØ detector are interpreted to try to identify and measure the kinematic properties of the particles that were produced in the collisions.

The three subsequent chapters discuss the actual analysis. The substantial efforts to accurately model single top quark production, and to identify, model and estimate the amount of the background processes are described in Chap. 5. The event selection, and the systematic uncertainties of the background estimation are also presented here. Chapter 6 explains how boosted decision trees can be constructed and used in experimental particle physics analyses to separate the detector footprints (events) of a physics process from the footprints of its background processes. Chapter 7 finally describes how the boosted decision trees used in this analysis are created and applied to the dataset, and how the single top cross section and the signal significance are measured. This chapter also presents cross checks of the results, and a direct measurement of the amplitude of the CKM matrix element $|V_{tb}|$—a fundamental parameter of the standard model that governs how top and b quarks mix.

References

1. D.J. Griffiths, *Introduction to Elementary Particles*, 2nd edn. (Wiley-VCH, Weinheim, 2008)
2. Wikipedia, The Free Encyclopedia (2010), http://en.wikipedia.org/wiki/Standard_Model, http://en.wikipedia.org/wiki/Particle_physics

Chapter 2
Theoretical Background

2.1 The Standard Model

2.1.1 Matter Particles

All known fundamental particles are classified as either quarks, leptons or gauge bosons. The quarks and leptons are spin-1/2 fermions and constitute the building blocks of matter. They are grouped into three generations, where the lightest particles are found in the first generation, and the heaviest in the third generation. Each generation contains a charged lepton, a charge-neutral neutrino, and an up-type and a down-type quark with charges $+2/3$ and $-1/3$, respectively. Quarks carry colour charge and never appear as free particles but in bound states called hadrons. The properties of the quarks and leptons are summarized in Table 2.1. For each particle in this table, there is an anti-particle with exactly the same mass, but opposite quantum numbers, such as electric charge and colour charge.

2.1.2 Particle Interactions

The quarks and leptons interact with each other via the exchange of spin-1 gauge bosons. There are three kinds of gauge bosons corresponding to the three inter-actions (forces) described in the standard model. The photon is the gauge boson for the electromagnetic interaction, which occurs between particles carrying electric charge. The massive W^+, W^- and Z^0 bosons mediate the weak force, and massless gluons are the carriers of the strong force acting on particles with colour charge. The standard model also predicts the existence of the Higgs mechanism, which generates the mass for the elementary particles. An overview of all particles and their interactions is illustrated in Fig. 2.1.

D. Gillberg, *Discovery of Single Top Quark Production*, Springer Theses,
DOI: 10.1007/978-1-4419-7799-1_2, © Springer Science+Business Media, LLC 2011

Table 2.1 Properties of the matter particles [1]

Generation	Quarks			Leptons		
	Flavour	Charge	Mass	Flavour	Charge	Mass
I	up u	+2/3	1.5–3.3 MeV	electron e	−1	0.511 MeV
	down d	−1/3	3.5–6.0 MeV	e neutrino ν_e	0	<2.2 eV
II	charm c	+2/3	1.27 GeV	muon μ	−1	105.7 MeV
	strange s	−1/3	70–120 MeV	μ neutrino ν_μ	0	<0.17 MeV
III	top t	+2/3	171.2 GeV	tau τ	−1	1.777 GeV
	bottom b	−1/3	4.2 GeV	τ neutrino ν_τ	0	<16 MeV

The masses for the u, d and s quarks are estimates of the "current quark mass" at a 2 GeV scale in the $\overline{MS}$ scheme, and the masses for the c and b quarks are the "running quark masses" using the same scheme. The top quark mass is from direct observations in data

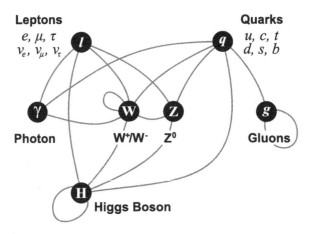

Fig. 2.1 Illustration showing all fundamental particles and interactions described in the standard model. The electromagnetic force is mediated by the photon that couples to all charged particles. The W and Z bosons carry the weak force between left-handed particles, and the gluon is the force carrier of the strong interaction, and couples to particles with colour charge. The Higgs boson, which is part of the standard model but not yet observed, couples to all massive particles

2.1.3 Gauge Theories

The standard model incorporates the gauge theories of the electroweak and strong interactions. A gauge theory is a quantum field theory (QFT) which is invariant under certain symmetry transformations. Massless gauge fields are introduced by demanding the Lagrangian for a gauge theory to be invariant under *gauge transformations*—symmetry transformations which depend on the space–time coordinate. Excitations (quanta) of a gauge field represent spin-1 gauge bosons that carry the force associated with the field.

The electroweak interaction belongs to the $SU(2)_L \times U(1)_Y$ gauge group. L here indicates that the weak force only couples to left-handed particles, and Y

refers to the weak hypercharge. In an unbroken form, the electroweak gauge group requires all of its bosons to be massless. This is not the case in nature since W and Z bosons are known to have large mass. In the standard model, particles acquire mass through the *Higgs mechanism*, which introduces a doublet of complex scalars whose self-interaction breaks the electroweak symmetry. This results in one physical scalar Higgs boson, which is the only elementary particle predicted by the standard model that is not yet observed.

The standard model is the combination of the electroweak and the strong interactions, which forms the $SU(3)_C \times SU(2)_L \times U(1)_Y$ gauge group. The first term in this expression is the gauge group for the strong force, and the subscript C here refers to the colour charge.

2.2 The Top Quark

2.2.1 Discovery

The top quark was predicted in 1977 when the b quark was discovered at Fermilab [2]. The b quark was observed to be a down-type quark, and since the theory requires each quark to have an isospin partner, the top quark was postulated as an up-type quark. It can be produced both via the strong interaction and via the electroweak interaction.

The top quark was discovered in 1995, 18 years after the b quark, by the DØ and CDF Collaborations at Fermilab [3, 4]. It was observed produced in pairs via the strong interaction. The Feynman diagrams for top pair production are shown in Fig. 2.2.

It took another 14 years before electroweak top quark production was discovered. This thesis presents this observation.

2.2.2 Properties

Just like the other up-type quarks, the top quark participates in both strong and electroweak interactions and has spin $1/2$ and charge $+2/3$e. However, it also exhibits several unique properties. It has the largest mass of any elementary particle; its mass is approximately that of a tungsten atom, nearly 40 times larger than the mass of the b quark and 10^4 times heavier than the up and down quarks. The large mass comes with a very short lifetime of about 0.4×10^{-24} s. This is shorter than the hadronization time scale of $\sim 3.0 \times 10^{-24}$ s, which means that the top quark decays as a free particle before undergoing fragmentation, transmitting its properties cleanly to the decay products.

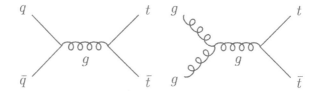

Fig. 2.2 Representative Feynman diagrams for $t\bar{t}$ production. The left diagram shows quark–antiquark annihilation, the right one shows gluon–gluon fusion. Quark–antiquark annihilation is the dominant production channel at the Tevatron (85, 15% for gluon fusion)

2.2.3 Decay

The top quark decays nearly exclusively to a W and a b quark. In the standard model, the branching ratio $\mathcal{B}(t \to Wb)$ is greater than 0.998. The b quark from the top decay will form a jet, but the W boson has many different decay modes. A top quark decay is therefore categorized by the decay of the W, which either decays to a lepton and a neutrino, or to a quark-antiquark pair. All lepton flavours are kinematically allowed, but the hadronic decay is limited to the first two generations of quarks since $m_W < m_t$. At leading order, a hadronic decay is three times as likely as a leptonic decay since quarks comes in three colours, and we get $\mathcal{B}(W \to q\bar{q}') = 1/3$ and $\mathcal{B}(W \to \ell v) = 1/9$ for a given quark/lepton flavour (except $W \to tb$ as mentioned above). Higher order corrections slightly alter this symmetry. A summary of the W decay modes is shown in Table 2.2.

2.3 Electroweak Top Quark Production

2.3.1 Introduction and Motivation

Electroweak top quark production is usually referred to as *single top* quark production since only one top quark is produced per event. Top quarks are here produced in charged current interactions via the *Wtb* vertex, which contributes by the factor

Table 2.2 Experimentally measured branching ratios for the decay of a real W [1]

Decay mode	Branching ratio (%)
$W^+ \to \ell v$	10.80 ± 0.09
$W^+ \to ev$	10.75 ± 0.13
$W^+ \to \mu v$	10.57 ± 0.15
$W^+ \to \tau v$	$11.25 \pm 0.20)$
$W^+ \to$ hadrons	67.60 ± 0.27

In the analysis presented in this thesis, the combined $\mathcal{B}(W \to \ell v)$ is used for each lepton flavour (see Table 5.4.)

$$\frac{-ig_w}{2\sqrt{2}}V_{tb}\gamma^{\mu}(1-\gamma^5) \tag{2.1}$$

to the matrix element for single top quark production. As a result, the single top quark production cross section is directly proportional to $|V_{tb}|^2$. From a measurement of the cross section, one can hence extract $|V_{tb}|$, without any assumptions on the number of generations in the standard model.

Single top quark production also offers opportunities to probe physics beyond the standard model, such as new exchange particles and flavour changing neutral currents. Further, after isolating single top quark events, it is possible to measure several of the top quark properties, such as the spin polarization. Finally, single top quark processes produce the same final state as the standard model Higgs boson process $WH \rightarrow Wb\bar{b}$ as well as the charged Higgs process $H^+ \rightarrow t\bar{b} \rightarrow Wb\bar{b}$. The background model, and essentially all analysis techniques developed for single top quark analyses, can hence also be used for Higgs searches.

Because of these interesting properties, single top quark production has been extensively studied, see for example Refs. [5–15].

2.3.2 Production Modes

At hadron colliders, there are three single top quark production modes, the s- and t-channel exchanges of a virtual W, and tW production. The next to leading order (NLO) cross sections for these processes at the Tevatron are given in Table 2.3.

t-channel single top quark production is the dominant single top quark production mode at the Tevatron. In this process, a virtual, space-like W boson ($Q_W^2 < 0$, where Q_W is the W four-momentum) interacts with a b quark from the proton sea. This process also has the alias Wg fusion, since the b quark originates from a gluon in the sea splitting into a $b\bar{b}$ pair. The most important Feynman diagrams for t-channel single top quark production are presented in Fig. 2.3. There is a $2 \rightarrow 3$ and a $2 \rightarrow 2$ diagram where the latter is a sub-process of the former where the gluon splitting in the sea has been ignored.

The s-channel single top quark production is, at leading order, the process $q\bar{q}' \rightarrow t\bar{b}$ which is illustrated in Fig. 2.4. In this process, the exchange particle is a time-like W boson with $Q_W^2 > (m_t + m_b)^2$.

Table 2.3 NLO single top quark production cross sections at the Tevatron (1.96 TeV $p\bar{p}$ collider) for $m_t = 170$ GeV [16]	Process	Cross section (pb)
	t-channel	2.34 ± 0.14
	s-channel	1.12 ± 0.06
	tW	0.28 ± 0.06

For comparison, the corresponding NLO $t\bar{t}$ production cross section is $7.91^{+0.61}_{-1.01}$ pb [17]

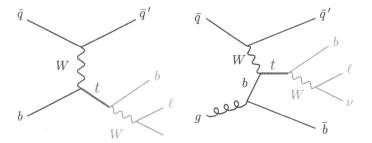

Fig. 2.3 The leading order $2 \rightarrow 2$ (*left*) and $2 \rightarrow 3$ process (*right*) Feynman diagrams for *t*-channel single top quark production. The left diagram is a subset of the right

Fig. 2.4 Leading order
s-channel single top quark
production

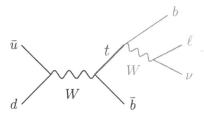

The *tW* process produces top quark with an on-shell W ($Q_W^2 = m_W^2$). The cross section for this process is very small at the Tevatron, and this production mode is therefore ignored in this analysis.

2.3.3 Measurement of $|V_{tb}|$

The Cabbibo–Kobayashi–Maskawa (CKM) quark mixing matrix describes the relationship between the quark mass eigenstates (d, s, b) and the weak eigenstates (d', s', b') during charge current interactions. Within the standard model with three generations, unitarity of the CKM matrix gives

$$|V_{ub}|^2 + |V_{cb}|^2 + |V_{tb}|^2 = 1. \tag{2.2}$$

Since $|V_{ub}|$ and $|V_{cb}|$ have been precisely measured, this implies a tight restriction on $|V_{tb}|$ [1]:

$$0.999090 < |V_{tb}| < 0.999177. \tag{2.3}$$

However, if we do not assume three generations, then Eq. 2.2 becomes

$$|V_{ub}|^2 + |V_{cb}|^2 + |V_{tb}|^2 + \cdots = 1, \tag{2.4}$$

and the constraints on $|V_{tb}|$ change to [15]:

$$0.06 < |V_{tb}| < 0.9994. \tag{2.5}$$

As previously mentioned, the single top quark production cross section is proportional to $|V_{tb}|^2$. From a measurement of the single top quark production cross section, we can therefore extract a measurement of $|V_{tb}|$. A measurement that differs significantly from the range specified in Eq. 2.3 would be clear evidence for physics beyond the standard model, and could possibly indicate the existence of a fourth generation of quarks.

The first direct measurement of $|V_{tb}|$ was presented in 2006 by the DØ Collaboration together with evidence for single top quark production [5, 6]. This analysis conducts a refined measurement using a larger dataset, see Sect. 7.5.

2.3.4 Single Top Kinematics

Figure 2.5 shows various kinematic distributions for the final state particles produced in s- and t-channel single top quark production (see Figs. 2.3, 2.4). These distributions are from the Monte Carlo samples used to simulate single top in this analysis. The modeling of these samples is described in Sect. 5.4.2

There are several characteristic kinematic features of single top quark production that can be seen in Fig. 2.5. The b quark emitted from the top quark decay tends to be central and has large transverse momentum. For the decay products of the W boson, we see that the lepton has a softer p_T spectrum than that of the neutrino. This occurs because the preferred direction of the lepton is anti-aligned with the top quark direction due to the $V-A$ nature of the weak force, as further discussed in Sect. 2.3.5. The b quark produced in association with the top quark in t-channel single top production tends to have high rapidity and low momentum and is often not reconstructed in the analysis. The light quark produced in the t-channel has reasonably large p_T, but its most distinguishing feature is the asymmetric $Q(\ell) \times \eta$ distribution, where $Q(\ell)$ is the charge of the lepton in the event. This asymmetry arises since the final state light quark produced during single t ($\bar{t}$) production most often is a d ($\bar{d}$) quark that moves in the same direction as the proton (antiproton) [15]. The light quark η will hence tend to have the same sign as the charge of the lepton from the top decay.

2.3.5 Polarization

As earlier discussed (Sect. 2.2.2), top quarks decay before they hadronize, transmitting their properties cleanly to the W boson and the b quark produced in the decay. In the standard model, the Wtb interaction is entirely left-handed, which means that single top quark production is a source of highly polarized top quarks [9].

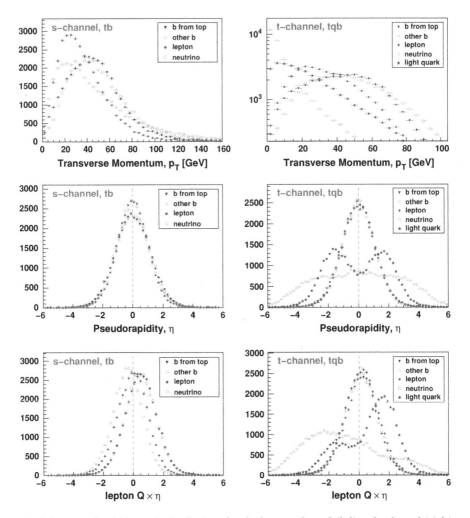

Fig. 2.5 Parton-level kinematic distributions for single top s-channel (*left*) and t-channel (*right*) from the single top Monte Carlo samples generated as described in Sect. 5.4.2. The p_T spectrum for each final state particle is shown in the top row, the corresponding η and $Q(\ell) \times \eta$ spectra are shown in the middle and bottom rows, respectively. These distributions were generated after parton showering was applied

The polarization of the top quarks becomes evident in the angular correlations between the decay products (see Fig. 2.6). The lepton is preferably emitted in the same direction as the top quark spin. The distribution of the angle θ_ℓ between the lepton momentum in the top rest frame, and the top polarization vector is given by [9, 15]:

$$F(\theta_\ell) = \frac{1}{2}(1 + \cos \theta_\ell). \tag{2.6}$$

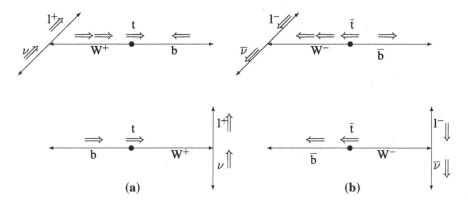

Fig. 2.6 Illustrations of spin and angular correlations in a top decay. *Double arrows* show the preferred direction of the spin and single lines represent the direction of the momentum in the rest frame of the parent particle. The top and anti-top quarks move to the left in all cases, and the preferred spin direction is against (towards) the direction of motion for the top (anti-top) since it is a left-handed (right-handed) particle. The two *upper diagrams* show a top (**a**) and an anti-top (**b**) decaying to a transversely polarized W, and the two lower diagrams show the corresponding decays to longitudinally polarized W bosons. In all cases, the charged lepton ℓ^+ (ℓ^-) tends to have its spin aligned with the spin of the top t $(\bar{t})$, and travel against the direction of the top t $(\bar{t})$. This results in a softened lepton momentum distribution as can be seen in Fig. 2.5. (Figure courtesy of [9])

In this analysis, several of the features introduced by top polarization are used to help identify single top quark events, see Sect. 7.1.2.

2.3.6 New Physics

Measuring the single top quark production cross section, and the different angular distributions is interesting as a test of the standard model, but also as a probe for several new physics scenarios beyond the standard model [15].

New physics can affect the single top quark production cross sections for the production modes (*tb*, *tqb* and *tW*) differently. The *s*-channel (*tb*) is most sensitive to new, heavy charged bosons. For instance the presence of a heavy W' boson, or a charged Higgs boson $H^{\pm}$, would increase the measured *s*-channel single top quark production cross section. The *t*-channel single top quark production would similarly be enhanced by flavour-changing neutral currents (FCNC). In the standard model, FCNC interactions are forbidden. Representative Feynman diagrams for single top production via an *s*-channel exchange of a heavy boson and a *t*-channel FCNC process are shown in Fig. 2.7.

Finally, physics beyond the standard model can alter the $V-A$ structure of the *Wtb* coupling. This would affect the top polarization, and hence also angular distributions such as $F(\theta_\ell)$ given in Eq. 2.6.

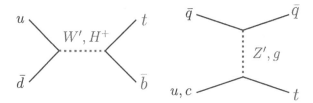

Fig. 2.7 *Left* representative Feynman diagram for anomalous single top quark production via the *s*-channel exchange of a heavy W' or charged Higgs boson. *Right* Diagram for flavour changing neutral current single top quark production via the *t*-channel

References

1. C. Amsler et al., Particle data group. Rev. Part. Phys. Phys. Lett. B **667**, 1 (2008)
2. S.W. Herb et al., Observation of a Dimuon resonance at 9.5 GeV in 400 GeV proton–nucleus collisions, Phys. Rev. Lett. **39**, 252 (1977)
3. S. Abachi et al., (DØ collaboration), Observation of the top quark. Phys. Rev. Lett. **74**, 2632 (1995)
4. F. Abe et al., (CDF Collaboration), Observation of top quark production in $p\bar{p}$ *Collisions*. Phys. Rev. Lett. **74**, 2626 (1995)
5. V.M. Abazov et al., (DØ Collaboration), Evidence for production of single top quarks and first direct measurement of $|V_{tb}|$. Phys. Rev. Lett. **98**, 181802 (2007)
6. V.M. Abazov et al., (DØ Collaboration), Evidence for production of single top quarks, Phys. Rev. D **78**, 012005 (2008)
7. T. Aaltonen et al., (CDF Collaboration), Measurement of the single-top-quark production cross section at CDF. Phys. Rev. Lett. **101**, 252001 (2008)
8. A. Heinson, et al., Single top quarks at the Fermilab Tevatron, Phys. Rev. D **56**, 3114–3128 (1997)
9. D. O'Neil, Performance of the HEC and the physics of electroweak top quark production at atlas, Ph.D. thesis, UMI-NQ-45351 (1999)
10. B. Abbott et al., (DØ Collaboration), Search for electroweak production of single top quarks in $p\bar{p}$ *Collisions*. Phys. Rev. D **63**, 031101 (2001)
11. V.M. Abazov et al., (DØ Collaboration), Search for single top quark production at DØ using neural networks. Phys. Lett. B **517**, 282 (2001)
12. D. Acosta et al., (CDF Collaboration), Optimized search for single top quark production at the Fermilab Tevatron. Phys. Rev. D **69**, 052003 (2004)
13. D. Acosta et al., (CDF Collaboration), Search for electroweak single-top-quark production in $p\bar{p}$ collisions at $\sqrt{s} = 1.96$ TeV. Phys. Rev. D **71**, 012005 (2005)
14. M.T. Bowen, et al., In search of lonely top quarks at the Tevatron, Phys. Rev. D **72**, 074016 (2005)
15. T.M.P. Tait, C.P. Yuan, Single top quark production as a window to physics beyond the standard model. Phys. Rev. D **63**, 1014018 (2000)
16. N. Kidonakis, Single top quark production at the Fermilab Tevatron: threshold resummation and finite-order soft gluon corrections Phys. Rev. D **74**, 114012 (2006)
17. N. Kidonakis, R. Vogt, Next-to-next-to-leading order soft gluon corrections in top quark hadroproduction. Phys. Rev. D **68**, 114014 (2003)

Chapter 3
Experimental Setup

In order to study the world's smallest particles, it is necessary to build the world's largest machines. This chapter presents an overview of the Tevatron, at present the world's highest energy collider, and the formidable DØ detector, in which the particle collisions are studied.

3.1 The Accelerator Chain

The Tevatron, situated at the Fermi National Accelerator Laboratory near Chicago, is currently the world's highest energy collider, with a centre of mass energy of 1.96 TeV. It is a circular, superconducting synchrotron in which protons (p) and anti-protons ($\bar{p}$) circulate in opposite directions and are brought together into collision in the B0 and D0 experimental areas. In these areas, two general purpose detectors, CDF and DØ respectively, measure the collision products.

An aerial view of Fermilab showing the accelerator facilities can be seen in Fig. 3.1. A 400 MeV hydrogen ion (H^-) beam is produced from hydrogen, accelerated by a Cockroft–Walton accelerator followed by a 165 m linear accelerator. The electrons are stripped off as the ions pass through a carbon fibre foil into the Booster synchrotron ring. Here the produced protons are accelerated to 8 GeV before being transferred to the Main Injector where the particles are accelerated to 150 GeV. The protons are arranged into a bunch structure and are delivered from the Main Injector to the Tevatron where the proton bunches are finally accelerated to 980 GeV.

Proton bunches from the Main Injector are also used to produce anti-protons. A proton beam of 120 GeV is directed at a nickel/copper target. The anti-protons produced are accelerated to 8 GeV and accumulated. Once the number of anti-protons is sufficiently large, the anti-protons are passed to the Main Injector where they are accelerated to 150 GeV for transfer into the Tevatron.

D. Gillberg, *Discovery of Single Top Quark Production*, Springer Theses,
DOI: 10.1007/978-1-4419-7799-1_3, © Springer Science+Business Media, LLC 2011

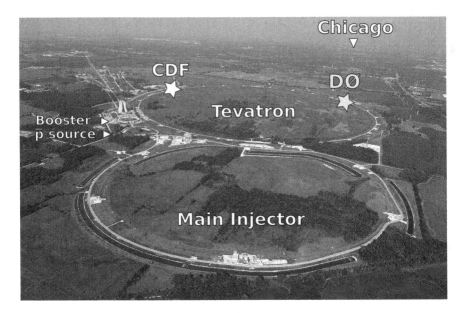

Fig. 3.1 Aerial view of Fermilab National Accelerator Laboratory showing some of the facilities described in Sect. 3.1, and the location of the DØ detector described in Sect. 3.2

Thirty-six bunches of protons and equally many bunches of anti-protons are delivered to the Tevatron with a 396 ns bunch spacing. The 36 bunches in each beam are organized into three super-bunches, separated by a 2 μs gap. The beams are focused at the collision points, and $p\bar{p}$ collisions occur during the bunch-crossings.

3.1.1 Luminosity and Cross Sections

In particle and nuclear physics, collision rates are measured in terms of the instantaneous luminosity, L. The rate of an arbitrary physics process X is given by

$$\frac{dN_X}{dt} = L\sigma_X \tag{3.1}$$

where σ_X is the cross section of the process. The cross section is an area, commonly expressed in picobarn (pb), where b $\equiv 10^{28}$ m^{-2}. It is usually desirable for a collider to provide a high instantaneous luminosity in order to achieve higher rates of rare processes (like single top quark production). The instantaneous luminosity often depends strongly on time. A more useful quantity in many cases is therefore the time independent *integrated luminosity*:

$$\mathcal{L} \equiv \int L dt. \tag{3.2}$$

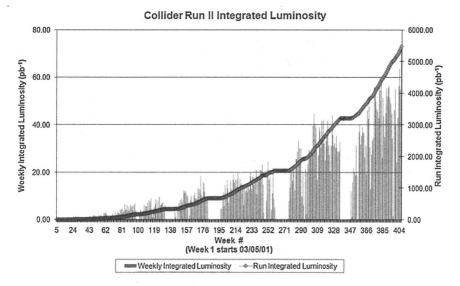

Fig. 3.2 Integrated luminosity $\mathcal{L}$ per week (*green bins*), and in total (*blue dots*), at Fermilab from May 2001 to December 2008. This analysis uses data from August 2002 to August 2007, approximately weeks 65–330 in the plot. The exact numbers are given in Table 5.1

The number of collisions which result in process X can now be expressed as

$$N_X = \mathcal{L}\sigma_X. \tag{3.3}$$

The integrated luminosity hence has the unit of inverse area, usually pb^{-1} or fb^{-1}. The instantaneous and integrated luminosities collected at Fermilab Run II are shown in Fig. 3.2. This analysis uses $2.3\,fb^{-1}$ of data ($\mathcal{L} = 2.3\,fb^{-1}$).

3.2 The DØ Detector

A sketch of the DØ detector [1] is shown in Fig. 3.3. The detector consists of four major subsystems. Starting from the interaction point and moving outward, these are: the central tracking system, the preshower detector, the calorimeters and the muon system.

3.2.1 The DØ Coordinate System

The DØ coordinate system is a right-handed Cartesian system with origin in the geometric centre of the detector. The x-axis lies in the horizontal plane pointing outwards from the centre of the Tevatron ring, the y-axis points straight up, and the z-axis is pointing along the beam pipe in the direction of the outgoing proton beam.

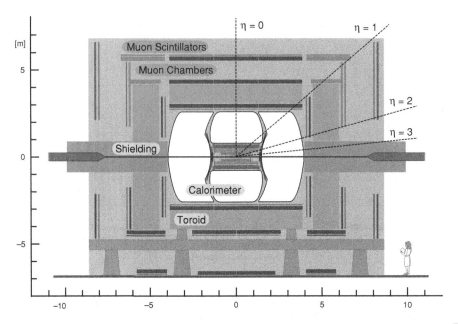

Fig. 3.3 A simplified cross section view of the DØ detector showing the different sub-detectors [1]

Since the protons and anti-protons are coming in along the z-axis, the (x,y)-plane is usually referred to as the transverse plane. The azimuthal angle ϕ for a particle is the angle between the positive x-axis and the transverse momentum vector $\vec{p}_T = (p_x, p_y)$ of the particle. The polar angle θ is the angle between $\vec{p}$ and the positive z-axis. However, since the collisions we want to study are boosted relative to each other along the z-axis, it is much preferred to use the rapidity y instead of the polar angle θ.

The rapidity for a particle is (in natural units) defined by

$$y = \frac{1}{2} \ln \frac{E + p_z}{E - p_z}. \tag{3.4}$$

This quantity is additive. A Lorentz boost β' along the z-axis is equivalent to a boost with rapidity $y' = \mathrm{arctanh}(\beta')$, and results in $y \to y + y'$. This means that differences in rapidity are invariant, and as a consequence, the shape of the high energy particle multiplicity spectrum dN/dy is also invariant under a boost along the z-axis. The energy, longitudinal momentum and velocity of a particle can be expressed in terms of rapidity as

$$E = \sqrt{p_T^2 + m^2}\cosh y, \quad p_z = \sqrt{p_T^2 + m^2}\sinh y, \quad \beta_z = \tanh y. \tag{3.5}$$

If the mass is small compared to the energy of the particle, $m \ll E$, then we can approximate the rapidity y with the pseudorapidity

$$\eta = \frac{1}{2} \ln \frac{|\vec{p}| + p_z}{|\vec{p}| - p_z} = -\ln\left(\tan\frac{\theta}{2}\right), \qquad (3.6)$$

and Eq. 3.5 becomes

$$E \approx |\vec{p}| = p_T \cosh\eta, \quad p_z = p_T \sinh\eta, \quad v_z \approx \frac{p_z}{|\vec{p}|} = \tanh\eta. \qquad (3.7)$$

As can be seen from the relations above, the pseudo-rapidity η is a purely angular variable.

A useful Lorentz invariant measure of the separation between two (massless) particles is

$$\Delta R = \sqrt{(\Delta\eta)^2 + (\Delta\phi)^2} \qquad (3.8)$$

where $\Delta\eta$ and $\Delta\phi$ are the separations between the particles in terms of η and ϕ, respectively.

3.2.2 The Central Tracking

The DØ central tracking system [1], illustrated in Fig. 3.4, consists of two tracking detectors: a silicon microstrip tracker (SMT) surrounded by the central scintillating fibre tracker (CFT). It is built inside a 2 T superconducting solenoid magnet with a

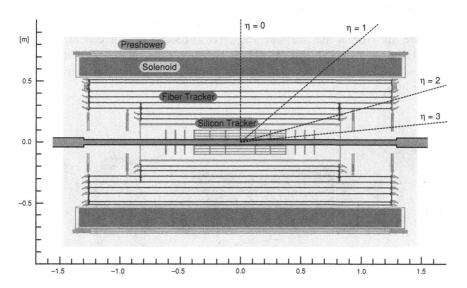

Fig. 3.4 Overview of the central tracking system at DØ during the Run IIa data taking period. During the summer of 2006, an additional silicon layer was added closer to the beam axis. The beam pipe was also replaced and the outermost silicon H-disks removed in the same upgrade

mean radius of 60 cm. This will bend the path of a charged particle in the $r-\phi$ plane, and from the radius of curvature the transverse momentum of the particle can be calculated according to

$$p_T = Brk, \tag{3.9}$$

where B is the magnetic field strength (2 T in this case), r is the radius of curvature, and k is the constant $0.3 \, \text{GeV}/(c\text{Tm})$.

In this analysis, the tracking system is used to identify and measure the momentum of electrons and muons, to determine the position of the primary interaction vertex, and to identify jets originating from b quarks. In order to perform these tasks accurately, it is important to have high spatial resolution. The following sections describe how the tracking systems have been designed in order allow for such precision measurements.

3.2.3 Silicon Microstrip Tracker

The basic unit of the SMT is called a wafer [1]. Silicon is the active material, and there is a voltage applied across the wafer. A charged particle passing through the wafer will create many electron–hole pairs that will drift across the unit and generate an electronic signal. The signal is amplified and read out in parallel microstrips arranged on one of the wafer surfaces. Two wafers can be arranged back-to-back with the microstrips on each side at a relative angle. This allows for stereo measurements of the particle hit.

An overview of the SMT with its barrel and disk structure is shown in Fig. 3.5. There are six barrels that measure the (r, ϕ, z)-coordinates of central (low η) tracks. There are also twelve so-called F-disks between, and at the end of the barrel segments, and four large "H-disks" in the forward region which can detect tracks with $2 < |\eta| < 3$.

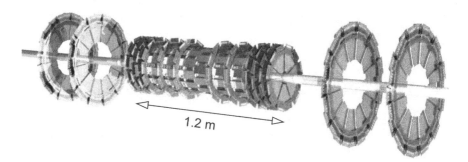

1.2 m

Fig. 3.5 The DØ Run IIa silicon microstrip tracker with its six barrel segments, twelve F-disks and four large H-disks. The two outermost H-disks were removed during the Run IIb upgrade

During the Run IIa run period (March 2001–March 2006), the barrel consisted of four double sided layers. During the Run IIb upgrade in spring of 2006, an additional layer was added inside the existing barrel [2]. To allow for this, the beam-pipe was replaced with a thinner one. The two most forward H-disks were removed due to radiation damage. The Run IIb data taking period started in August 2006.

3.2.4 Central Fibre Tracker

The CFT [1] consists of eight concentric cylinders that enclose the SMT (see Fig. 3.4). The cylinder walls are made of two layers of closely spaced scintillating fibres. The fibres in one of the two layers are aligned with the beam axis, while the fibres in the other layer are arranged at a three degree relative angle allowing for stereo measurements to be made.

When a charged particle passes through a scintillating fibre, a small fraction of its energy may excite molecules in the material that will emit visible light during the subsequent deexcitation. The photons will travel through the fibre and be collected in "visible light photon counters" (VLPCs) outside the detector.

3.2.5 Preshower Detectors

The central and forward preshower detectors consist of lead radiators combined with scintillating material, and are placed in front of the calorimeters. They are designed to identify and measure the energy of particles that interact with matter before reaching the calorimeters. This aids the identification of electrons and photons, since they often start to shower in the solenoid magnet, which alone accounts for about one interaction length of material in the central region.

3.2.6 The DØ Calorimeters

The DØ calorimeters [1, 3] are used to identify and measure the energy and direction of electrons, photons, jets, muons and missing transverse energy $\not{E}_T$, and are hence crucial for this analysis. There are three cryostats with nearly equal size, the central calorimeter and the two endcap calorimeters (Fig. 3.6). Each calorimeter is divided in layers: innermost there are four electromagnetic (EM) layers, followed by the fine and coarse hadronic layers.

The design of the EM layers is optimized for measurement of EM showers produced by electrons and photons. The third EM layer has increased granularity

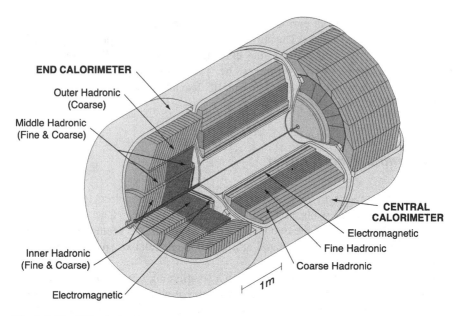

END CALORIMETER

Outer Hadronic
(Coarse)

Middle Hadronic
(Fine & Coarse)

**CENTRAL
CALORIMETER**

Electromagnetic

Fine Hadronic

Inner Hadronic
(Fine & Coarse)

Coarse Hadronic

Electromagnetic

⊢ 1m ⊣

Fig. 3.6 The DØ calorimeters

since this is where maximum shower development is expected. Most EM showers will not penetrate into the hadronic calorimeter, which is designed for good measurement of hadronic showers. Muons only deposit a small amount of energy in the calorimeter, and neutrinos no energy at all. Some energy will also be deposited in poorly instrumented regions and hence give no or little signal. This absence of measured energy results in a momentum imbalance in the transverse plane. This imbalance is called the missing transverse energy, $\not{E}_T$.

The basic unit of the DØ calorimeters is a calorimeter cell. Such a cell consists of an absorber plate (U, Cu or Fe) followed by a gap filled with liquid argon. In the middle of this gap is a G-10 board, with a 2.0–2.5 kV potential with respect to the grounded absorber plate. This potential difference induces a drift field across the liquid argon. As an incoming particle interacts with the dense matter in the absorber plate, a shower of secondary particles is produced. As they pass through the liquid argon, they ionize argon atoms, and negative charge will drift towards the signal boards. This results in a signal proportional to the energy loss of the incoming particle. A schematic view of two typical calorimeter unit cells is given in Fig. 3.7. Several unit cells stacked on top of each other are read out together.

Figure 3.8 shows a side view of the calorimeters. We can see the layer structure, but also that cells with the same η (and ϕ) are arranged in "pointing towers", i.e., the towers point towards the centre of the detector (the interaction point). Cells have a size of about $\Delta\eta \times \Delta\phi = 0.1 \times 0.1$ except in the third EM layer where the granularity is doubled.

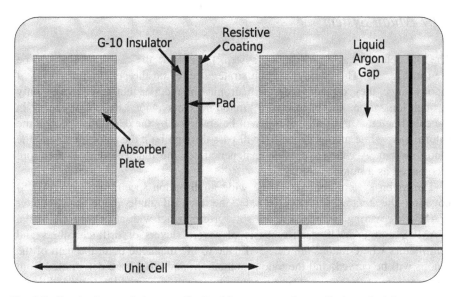

Fig. 3.7 Sketch of two calorimeter cells. Particles penetrate these cells from the left

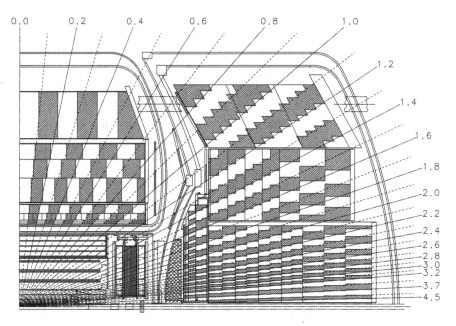

Fig. 3.8 Side view of a quarter of the DØ calorimeters. The *lines* with *numbers* are lines of constant η. Cells are arranged into pointing towers along these lines. There are four electromagnetic layers in all cryostats, three fine hadronic layers in the central cryostat and four in the end caps, and one coarse hadronic layer in the central cryostat and three in the end caps

3.2.7 Muon System

The DØ detector has a large muon system [1] outside the calorimeter as can be seen in Fig. 3.3. The muon detection strategy relies on the penetration power of muons since they do not undergo hadronic interactions but lose energy only through ionization. A typical high p_T muon deposits about 1.8 GeV of energy in the calorimeter. Almost all hadrons will be absorbed by the dense materials in the calorimeter, while muons generally will pass through both the calorimeter and the muon system. A charged particle that penetrates the muon system is therefore recognized as a muon.

The muon system consists of the wide angle muon spectrometer (WAMUS) covering the central detector ($|\eta| < 1$), the forward angle muon spectrometer (FAMUS) covering $1 < |\eta| < 2$ and a solid-iron magnet with at field of 1.8 Tesla. WAMUS and FAMUS each consists of several layers of drift chambers and scintillators where muons are detected. Due to the magnetic field, the path of the muons will be curved, and the muon momentum and charge are determined from the curvature of the tracks. These measurements are improved by using additional information from the central tracking system and the calorimeters.

3.2.8 Triggers

The collision rate at the Tevatron is 2.5 MHz, i.e., 2.5 million events per second. To read out all detector signals produced by one event requires 250 kB of data. There is no practical technology available to collect and store data at this rate since most events produced are uninteresting events. Production of heavy mass resonances, such as W and Z bosons or top quarks, occur at a much lower rate.

The DØ trigger system is designed to operate in this very high collision rate environment. It is organized into three major levels:

Level 1: This trigger level is required to reduce the event rate from 2.5 MHz to 1.4 kHz. The trigger is hardware based, and receives input from the calorimeter, the muon system and the luminosity system.

Level 2: The hardware Level 2 trigger has two stages and reduces the rate to 1 kHz. The first stage consists of several "preprocessors" that each receive information from one of the individual subdetectors to produce objects such as tracks, electrons, jets and muons. The second stage makes a trigger decision based on the properties of these objects.

Level 3: The final trigger level relies entirely on software that is run on a dedicated computer farm. The trigger has about 100 ms to make its decision, and reduces the rate to 50 events per second. There are algorithms performing close-to-offline reconstruction of electrons, muons, jets and missing E_T. Events satisfying this trigger are stored and transferred to full offline reconstruction.

References

1. V.M. Abazov et al., The upgraded D0 detector, Nucl. Instrum. Meth. **A565** 463–537 (2006)
2. J. Fast (DØ Run IIb Silicon Group), The DØ silicon detector for Run IIb at the Tevatron Nucl. Phys. B **125** (2003)
3. J. Kotcher, Design, performance and upgrade of the DØ calorimeter, FERMILAB-CONF-95-007-E (1995)

Chapter 4
Event Reconstruction

This chapter describes how the detector subsystems are used in order to identify the physics objects, such as jets and electrons, which are created from a hard scatter $p\bar{p}$ collision. Two aspects of the event reconstruction are discussed: object reconstruction and object identification.

Object reconstruction starts with converting the raw detector signals to "hits" with a corresponding position and measured energy. The hits are next clustered depending on their position to form a basic physics object, meaning either a track or a calorimeter energy cluster. From these basic physics objects, the final physics objects are created, which in this analysis are: electrons, muons, vertices, jets and $\not{E}_T$. For illustrations of hits and reconstructed physics objects, see Appendix A.

During object identification, quality requirements are applied to each object. The following sections describe how the physics objects used in this analysis are reconstructed, and what object identification requirements are applied.

4.1 Tracks

Tracks are used to reconstruct many of the physics objects used in this analysis, namely electrons, muons, the primary vertex and b jets. As a charged particle traverses the tracking system, its path is bent by the magnetic field of the solenoid, and small amounts of energy are deposited along the particle trajectory in many of the tracking layers. The DØ tracking algorithm reconstructs particle tracks from such hits. This is not an easy task since any given event contains thousands of hits, and not all of them are from the hard scatter collision, but also from secondary collisions and electronic noise.

The DØ track reconstruction first constructs a list of track candidates using two different methods. The histogram track finding method (HTF) [1] is based on a Hough transformation which originally was used to find patterns in pictures taken in bubble chambers [2]. All possible combinations of two hits are created, and for each such combination, the angular direction and the curvature ρ for the trajectory from the beam axis through both hits are calculated. These quantities are filled in two

D. Gillberg, *Discovery of Single Top Quark Production*, Springer Theses,
DOI: 10.1007/978-1-4419-7799-1_4, © Springer Science+Business Media, LLC 2011

dimensional histograms, and a peak is formed for a track, since the track, and also all the pairs of hits of the track, have the same direction and curvature. Fake track segments, created from electronic noise, are spread uniformly in these histograms.

The alternative algorithm (AA) [3] creates track seeds from hits in the silicon tracker and forms roads. Hits along those roads in additional tracking detector layers are added to the track if they improve the overall χ^2 of the track fit. Compared to the histogramming method, this method has a better efficiency for low p_T tracks, and tracks from secondary vertices. It is also less susceptible to fake tracks.

Finally, the tracks provided by these two methods are used as input to the global track reconstruction (GTR). The tracks are here created, combined, refitted and smoothed using a Kalman filter algorithm [4], resulting in the final set of tracks in the event.

4.2 Primary Vertices

A precise determination of the primary interaction point along the beam axis is important for determining the direction of jets, muons and electrons, and also for identifying secondary vertices, which is crucial for b tagging. The location of the primary vertex is close to the geometrical centre of the detector in the (x, y)-plane, but the z position can vary over roughly one metre along the beam axis from event to event.

The primary vertices of an event are reconstructed by means of an adaptive primary vertex algorithm [5]. This algorithm first determines the beam position in the (x, y)-plane and its width from a χ^2 fit of all tracks. The beam axis is next divided into segments of length 2 cm. The tracks with $p_T > 0.5$ GeV, and at least two SMT hits that are pointing back to a given segment are clustered. The tracks in each cluster are fitted to a common vertex using the Kalman filter technique [4]. After the initial vertex fitting, the tracks with the highest contribution to the vertex χ^2 (the "worst" tracks) are removed, and the vertex is refitted, until the total $\chi^2 < 10$.

The final vertex is calculated from the remaining tracks. In the case where more than one vertex is found, the p_T distributions of the tracks associated with each vertex are used to define a probability that each track originated at the particular vertex [5]. The vertex with the lowest probability of being a minimum bias vertex is selected as the hard scatter vertex.

4.3 Calorimeter Clusters

Before using the measured energies in the calorimeter for object reconstruction, it is necessary to suppress noise. The procedure to deal with hot cells (cells that give a high measured energy due to hardware problems), and energy mis-measurements due to electronic noise, are briefly discussed below.

Each calorimeter cell is considered a massless object, and is assigned the four vector $(E_{cell}, \vec{p}_{cell})$, where E_{cell} is the measured energy and $\vec{p}_{cell}$ is a vector of magnitude $|E_{cell}|$ directed from the primary vertex to the centre of the cell.

Starting from the list of all calorimeter cells, the following selection criteria are applied:

a. Cell are required to fulfil $|E_{cell}| > 2.5\sigma_{cell}$, where σ_{cell} is the measured energy width due to electronic noise.
b. Cells identified as hot cells by the NADA algorithm [6] are removed.
c. According to the T42 algorithm [7], all cells with $E_{cell} > 4\sigma_{cell}$ are first selected. Next, cells with $E_{cell} > 2\sigma_{cell}$ are selected if they have a neighbouring cell with $E_{cell} > 4\sigma_{cell}$. All other cells are removed.

Only the cells surviving this selection are used to reconstruct the calorimeter showers. For computing time reasons, cells belonging to a given calorimeter tower are first combined into a tower object. Each tower points to the geometrical centre of the detector, and contains both electromagnetic and hadronic layers as can be seen in Fig. 3.8. The four momentum of a tower object (or any other cluster of cells or towers, such as a jet), is defined by the four vector sum of the cells.

4.4 Electrons

The characteristic signature of an electron is a track in the inner tracking system, and a narrow and short shower in the electromagnetic section of the calorimeter. Electrons are hence reconstructed using information from both the calorimeter and the central tracker.

Electromagnetic clusters (EM clusters) are reconstructed by merging calorimeter tower objects (Sect. 4.3) using a simple cone algorithm [8]. Only the energy deposited in the electromagnetic part of the calorimeter is considered by the algorithm. Towers with $E_T > 1.5$ GeV are used as seeds, and an EM cluster is created by including the towers in a radius of $\Delta R = 0.2$ (see Eq. 3.8).

The following variables are used in this analysis to identify and assess the quality of an EM cluster:

Electromagnetic Fraction, $f_{EM} = E_{EM}/E_{total}$: This is the ratio of the energy E_{EM} deposited in the electromagnetic layers over the total cluster energy E_{total}, which includes the hadronic layers. For an electron, this fraction is expected to be close to one.

Isolation, f_{iso}: The isolation of an EM cluster is defined by

$$f_{iso} = \frac{E_{total}(\Delta R < 0.4) - E_{EM}(\Delta R < 0.2)}{E_{EM}(\Delta R < 0.2)}. \tag{4.1}$$

$E_{total}(\Delta R < 0.4)$ is the energy in a cone of radius $\Delta R = 0.4$ around the EM cluster. For a real, isolated electron, this energy should not be much larger than the central electromagnetic energy $E_{EM}(\Delta R < 0.2)$.

H-Matrix χ^2: This quantity is constructed from seven variables that describe the longitudinal and transverse shower shape of the EM cluster, namely: the energy deposited in each of the four EM layers, $\log_{10}(E_{EM})$, the primary vertex z position, and transverse shower width in the third EM layer. A 7×7 covariance matrix is constructed using these shower shape variables for simulated electrons that have been accurately modelled to agree with observed shower shapes of test beam electrons [9]. Using this matrix, χ^2_{HM} can be calculated for any given EM cluster.

Track Match χ^2: This is the χ^2 of the fit of the closest track with the centre of the EM cluster. It can be converted to a probability for the track to be associated with the EM cluster, $P(\chi^2)$, which is what is used in this analysis.

Likelihood $\mathcal{L}_{EM}$: The electron likelihood [10] is defined such that real electrons tend to have values close to 1, while fakes tend to have values close to 0. It only applies to track-matched electrons and is based on seven variables including both calorimeter and tracking information.

The electron definitions used in this analysis are the following:

Ultraloose Electron: An ultraloose electron is required to have $f_{EM} > 0.9$, $\chi^2_{HM} < 50$, $f_{iso} < 0.15$ and $p_T > 15$ GeV. There are no requirements for a matching track. This electron definition is used for modeling the multijet background, see Sect. 5.4.6.

Loose Isolated Electron: In addition to the ultraloose requirements, a loose isolated electron must have a track match with a non-zero χ^2 probability: $P(\chi^2) > 0$. The matching track is required to have $p_T > 5$ GeV and be pointing back close to the primary vertex: $\Delta z(\text{track}, \text{PV}) < 1$ cm.

Tight Isolated Electron: A tight isolated electron must pass all the loose isolated electron requirements and in addition have $\mathcal{L}_{EM} > 0.85$.

4.5 Jets

A quark or gluon emitted from the hard scatter collision will undergo a complicated process that results in a *jet*-a spray of hadrons with a total momentum close to the momentum of the emitted parton (Fig. 4.1). Jets vary widely in shape and particle content, and deposit energy both in the electromagnetic and hadronic layers of the calorimeter. All single top events produce two or more jets. Accurate knowledge of the jet energies and their directions is therefore very important.

The jets used in this analysis are reconstructed using the Run II Improved Legacy Cone Algorithm [11, 12] with cone size $\mathcal{R} = 0.5$. The calorimeter tower objects, created as described in Sect. 4.3, are first combined into preclusters of radius 0.3 using the simple cone algorithm (same algorithm as for EM clusters). It is ensured that no preclusters share any towers. The centre of each precluster, but also the midpoint between any pair of preclusters, are used as seeds for the final jet reconstruction algorithm.

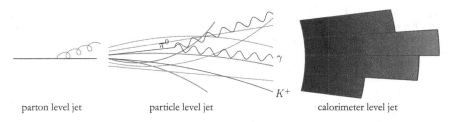

Fig. 4.1 Illustration of the evolution of a jet. A parton jet, consisting of a quark and a radiated gluon (*left*), hadronizes and forms a particle jet (*middle*) that creates electromagnetic and hadronic showers in the calorimeter. The energy of these showers is measured in the calorimeter cells, which are organized into pointing towers, and a jet object is reconstructed from these towers (*right*)

For each such seed, a jet is created by including all towers within a cone of size $\mathcal{R}$ around the seed. The centre axis of the jet is calculated, and the jet is redefined as the combination of towers within $\Delta R < \mathcal{R}$ of the new midpoint. This is repeated recursively until a stable cone is found. In the final step of the algorithm, overlaps between jets are removed. Two jets are merged if the shared energy is more than 50% of the energy of the sub-leading jet. If not, each shared tower is assigned to the jet closest in (y, ϕ)-space.

When measuring a jet in the calorimeter, there might be large fluctuations due to finite energy resolution and calorimeter cell granularity. The measured energy will on average be lower than the true energy since hadronic showers have a lower calorimeter response compared to electromagnetic showers, and since some particles of the jet may pass through uninstrumented regions. To account for these effects, jets are corrected by the jet energy scale (JES) [15] according to:

$$E_{\text{jet}}^{\text{corr}} = \frac{E_{\text{jet}}^{\text{raw}} - O}{R_{\text{jet}} F_\eta S}. \tag{4.2}$$

The components of Eq. 4.2 are described below.

Uncorrected Jet Energy $E_{\text{jet}}^{\text{raw}}$: The measured energy of all cells in the jet.

Offset Energy O: The energy not associated with the hard scatter. The main sources for this energy are energy deposited from jets produced in additional "min-bias" interactions, and energy due to electronic noise. This correction is shown in Fig. 4.2.

Inter-Calibration F_η: This is a calibration factor applied to make the response uniform as a function of jet η across the central and end-cap calorimeters and the inter cryostat regions. The size of this correction for a typical jet in this analysis is around 5%.

Jet Response R_{jet}: This is the main JES correction. The jet response in the DØ calorimeters is significantly lower than unity for several reasons: hadronic showers have a lower calorimeter response than electromagnetic showers; energy is lost in material in front of the calorimeter, such as tracking material and the solenoid

Fig. 4.2 The jet offset correction as a function of jet η. There are different curves depending on the number of primary vertices in the event, which are created from additional interactions. The correction is quite large for forward jets when there are additional $p\bar{p}$ interactions in the event

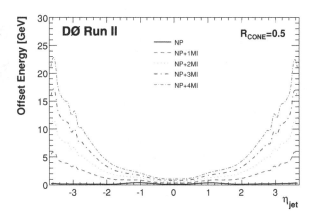

magnet; some particles in the jet might escape undetected, for instance due to uninstrumented regions or since they are neutrinos. The magnitude of this correction is shown in Fig. 4.3.

Showering Correction S: The DØ Run II jet algorithm reconstructs the jet from the deposited energy within the jet cone. Due to effects like shower development in the calorimeter and magnetic field bending, there will be energy leaving and entering the jet cone. The showering correction S corrects for the net energy difference due to such showering effects.

Fig. 4.3 *Top*: jet response, measured in γ + jet events, as a function of the jet energy estimator $E' = p_T(\gamma)\cosh\eta_{\text{jet}}$, which approximates the particle level jet energy. *Bottom*: the difference between the measurements and the parametrized jet response function, and the uncertainty band from the fit

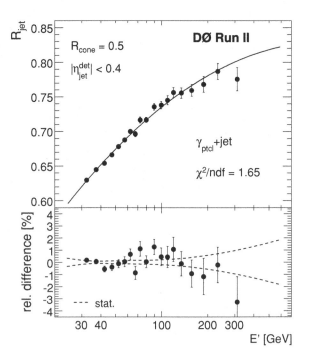

The jets used in this analysis are JES corrected as described above. They are also required to fulfil a set of selection criteria recommended by the DØ Jet-ID Algorithm Group [13, 14]. These criteria include requirements on the fraction of the jet energy in the outermost, coarse hadronic layer, f_{CH}, the fraction of the energy in the electromagnetic layers f_{EM}, and a trigger level 1 ratio requirement. In addition to these identification criteria, this analysis requires all jets to have $p_T > 15$ GeV, $|\eta| < 3.4$, and not to overlap with any loose isolated electron.

4.6 Muons

The starting point for muon reconstruction is the formation of a track from hits in each layer of the muon system. The track is combined with an existing track in the central tracking system reconstructed as described in Sect. 4.1. This greatly improves the p_T resolution compared with only using the muon system.

The following muon definitions are used in this analysis:

Loose Isolated Muon: A loose muon is required to have tracks with hits both in the drift tubes and the scintillators, and in two of the three detector layers of the muon system outside the toroid. A loose cosmic ray rejection timing requirement is applied, and the track reconstructed in the muon system must match a track reconstructed in the central tracker that has at least two hits in the silicon tracker. The χ^2 for the match between the two tracks must be less than 4. The muon track is required to be close to the primary vertex: $z(\text{track}, \text{PV}) < 1$ cm and it must not be overlapping with any jet in the event: $\Delta R(\mu, \text{jet}) > 0.5$.

Tight Isolated Muon: Tight isolated muons fulfil the loose muon requirements, and in addition the following isolation criteria:

a. the momenta of all tracks in a cone of radius $R < 0.5$ around the muon direction, except the track matched to the muon, add up to less than 20% of the muon p_T

b. the energy deposited in the calorimeter around the muon trajectory in the range $0.1 < \Delta R < 0.4$ must be less that 20% of the muon p_T.

4.7 *b* Jets

b jets are jets originating from the hadronization of *b* quarks. These objects are particularly important in this analysis since single top quark events produce two *b* quarks in the final state. As the *b* quark hadronizes, a *B* hadron will be formed, which is a bound state of a *b* quark and one or two light quarks. *B* hadrons have significantly longer lifetimes than lighter hadrons, and typically travel a few millimetres before decaying. As a consequence, *b* jets will usually have a decay vertex displaced from the primary interaction point that can be reconstructed as a secondary vertex. Another distinguishing property is that about 20% of all *b* jets contain a muon inside the jet cone. These features, and other kinematic properties, can be used to distinguish heavy flavour jets from (ordinary) light flavour jets.

This analysis uses a neural network (NN) b jet tagger designed by DØ 's B-ID Group to identify b jets [16]. Jets are first required to be "taggable", meaning that there are at least two good tracks associated with the jet such that a secondary vertex can be constructed for every jet. Taggable jets are then "tagged" by the tagging algorithm.

The NN tagger uses seven variables to discriminate b jets from other jets. The most important variable is the decay length significance of the secondary vertex, defined as the distance from the primary to the secondary vertex divided by the uncertainty of this quantity. The other variables are: the invariant mass of all tracks associated with the secondary vertex (SV); the χ^2 per degree of freedom for the reconstruction of the SV from the tracks; the number of tracks pointing to the SV; the number of SVs associated with the jet; and the probability that the jet tracks originate from the PV calculated from the minimal distance between each of the jet tracks and the PV. The NN tagger assigns an output value between 0 and 1 proportional to the probability that the jet is a b jet. Only jets with $|\eta| < 2.5$ are considered by the algorithm.

There are several operating points defined for the NN tagger. This analysis uses the TIGHT and the OLDLOOSE operating points, where the TIGHT means an NN output greater than 0.775, and OLDLOOSE means NN output greater than 0.5. Each event is required to have either one jet fulfilling the TIGHT NN b-tagging quality, or to have two jets tagged by the OLDLOOSE operating point. The average fake rates for the TIGHT/OLDLOOSE operating points are 0.82/2.5% for data jets in the central calorimeter, and their average b-tagging efficiencies on data are 49/61% for jets with $|\eta| < 2.5$.

4.8 Missing Transverse Energy, $\not{E}_T$

All single top events considered in this analysis have a high p_T neutrino in the final state. Neutrinos interact very weakly with matter, and their energy and momentum cannot be directly measured. However, since momentum is conserved, one can indirectly measure the p_T of the neutrino from the momentum imbalance in the transverse plane. This imbalance is called the missing transverse energy, $\not{E}_T$, and is defined by the negative sum of the transverse momenta of all particles observed in the detector. In practice, the (uncorrected) missing transverse energy is calculated by

$$\vec{\not{E}}_T = -\sum_{i}^{N_{\text{cells}}} \vec{p}_{Ti}.$$

(4.3)

where p_{Ti} is the transverse momentum for cell i (see Sect. 4.3 for the p_T definition for a calorimeter cell). Only cells in the electromagnetic and fine hadronic layers of the calorimeter are included since the energy resolution is poor in the coarse hadronic layers.

The missing energy defined in Eq. 4.3 needs to be corrected if there are reconstructed muons in the event, and due to energy corrections of jets, electrons and photons. A muon only deposits a small amount of energy in the calorimeter. If a loose isolated muon is present in the event, $\vec{\not{E}}_T$ is corrected by subtracting the component of the muon momentum that was not detected in the calorimeter. The same principle applies for jets. The momentum component added due to jet energy scale for each jets needs to be subtracted from the raw $\not{E}_T$. There are also small corrections needed if there are electrons or photons in the event due to the electron and photon energy scales.

References

1. A. Khanov, HTF: histogramming method for finding tracks. The algorithm description, DØ Note 3778 (2000)
2. P. V. C. Hough, Machine analysis of bubble chamber pictures, International Conference on High Energy Accelerators and Instrumentation (L. Kowarski, ed.), pp. 554–556 (1959)
3. D. Adams, Finding Tracks, DØ Note 2958 (1998)
4. H. Greenlee, The DØ Kalman track fit (2004). DØ Note 4303
5. A. Schwartzman and C. Tully, Primary Vertex Reconstruction by Means of Adaptive Vertex Fitting, DØ Note 4918 (2005)
6. B. Olivier, U. Bassler, G. Bernardi, NADA: A New Event by Event Hot Cell Killer, DØ Note 3687 (2000)
7. J.-R. Vlimant et al., Technical Description of the T42 Algorithm for Calorimeter Noise Suppression, DØ Note 4146 (2003)
8. J. Hays et al., Single Electron Efficiencies in p17 Data and Monte-Carlo, DØ Note 5025 (2006)
9. M. Narain, Electron identification in the D0 detector, Fermilab Meeting: DPF 92, Batavia, FERMILAB-CONF-93-054-E. (1992)
10. J. Kozminski et al., The Electron Likelihood in p14, DØ Note 4449 (2003)
11. Gerald Blazey, Run II Jet Physics, DØ Note 3750 (2000)
12. E. Bustato, B. Andreui, Jet algorithms in the DØ Run II Software: Description and User's Guide, DØ Note 4457 (2004)
13. A. Harel, Jet ID Optimization, DØ Note 4919 (2006)
14. A. Harel and J. Kvita, p20 JetID Efficiencies and Scale Factors, DØ Note 5634 (2008)
15. A. Juste et al. (DØ JES Group), Jet Energy Scale Determination at DØ Run II, DØ Note 5382 (2007)
16. M. Anastasoaie, S. Robinson, and T. Scanlon, Performance of the NN b-Tagging Tool on p17 Data, DØ Note 5213 (2006)

Chapter 5
Analysis: Event Selection

Single top quark production is a very rare process relative to its major backgrounds. The background arises from several distinct sources, each mimicking the single top signal in its own way. In essence, single top is kinematically "wedged" between W+jets and $t\bar{t}$ backgrounds, and there is no easy way to reduce these backgrounds simultaneously. Instead, each background needs to be probed for its individual distinguishing features. In order to identify these, and to correctly evaluate the amounts of signal and background in the dataset, it is necessary to create an accurate signal and background model.

This chapter explains the analysis strategy and describes the dataset used, the selection criteria applied, and the momentous task of modeling the signal and all background processes in the data.

5.1 Strategy

As explained in Sect. 2.2.3, single top quarks decay to a W boson and a b quark almost 100% of the time. W bosons further decay leptonically or into jets. This analysis focuses on single top decays in the electron and muon channels.

The composition of the background components is quite different for events with different jet multiplicities and lepton flavours. The data are therefore divided into many *channels* (orthogonal samples) depending on the lepton flavour, the number of reconstructed jets, and the number of b-tagged jets. The analysis is optimized individually in each such channel.

The general event selection strategy is to maximize signal acceptance by using a loose event selection and thereafter use a multivariate technique (in this case boosted decision trees) to separate signal from background.

D. Gillberg, *Discovery of Single Top Quark Production*, Springer Theses,
DOI: 10.1007/978-1-4419-7799-1_5, © Springer Science+Business Media, LLC 2011

Table 5.1 Integrated luminosities of the datasets used in this analysis (also, see Fig. 3.2)

Channel	Trigger version	Delivered	Recorded	Good quality
Integrated luminosity (pb^{-1})				
Run IIa electron	v8.00–v14.93	1,312	1,206	1,043
Run IIa muon	v8.00–v14.93	1,349	1,240	1,055
Run IIb e and mu	v15.00–v15.80	1,497	1,343	1,216
Total Run II integrated luminosity				2.3 fb^{-1}

5.2 Data Set

The data sample was collected between August 2002 and August 2007 during the Run IIa and Run IIb run periods. The Run IIb data were recorded at higher instantaneous luminosities, and with the upgraded detector as described in Sect. 3.2. The integrated luminosity for the dataset can be seen in Table 5.1.

Each electron data event is required to satisfy at least one trigger in a list of several hundred photon, electron, jet and e+jets triggers. For muons events, a similar list is used containing jet, muon and μ+jets triggers. Studies show that essentially all events that pass the event selection are accepted by these trigger requirements. The trigger efficiency used for the background modeling is 100%, with an uncertainty of 5–10% as discussed in Sect. 5.8.

5.3 Background Processes

The three major backgrounds for single top are W+jets, $t\bar{t}$ and multijet production. W+jets is the largest background for events with two jets, and $t\bar{t}$ is the largest background for events with four jets. There are also backgrounds from Z+jets and diboson processes.

Figure 5.1 shows example Feynman diagrams with the event signature particles highlighted for single top and the major background processes.

W+jets: W+jets events produce an on-shell W boson and one or several jets. The $Wb\bar{b}$ subprocess ($p\bar{p} \rightarrow Wb\bar{b} + X$) has the same final state as single top: two real b quarks and a W. The Wjj, $Wc\bar{c}$ and Wcj subprocesses, where j refers to a light jet, enter the data when jets are misidentified as b jets.

Top Pair Production: $t\bar{t}$ events produce two on-shell top quarks. $t\bar{t} \rightarrow \ell$+jets events have two b jets and $W \rightarrow \ell\nu$, just as single top, but have in addition two high p_T jets. This $t\bar{t}$ decay channel constitutes a large background for the high jet multiplicity channels. Dilepton events have an extra $W \rightarrow \ell\nu$ in the final state and make their way into the dataset when one of the leptons is not reconstructed.

Multijet: There is an instrumental background from multijet events in which one jet fakes an isolated lepton and imprecise jet calibration induces false $\not{E}_T$.

Single Top Production

W + jets

Top Pair Production

Multijets   ### Diboson

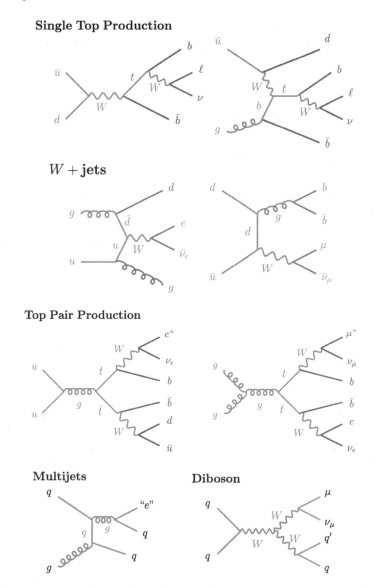

Fig. 5.1 Representative diagrams for single top and the major background processes. The "*e*" in the multijet diagram illustrates a quark that is mis-identified as an electron

Z+jets: Z+jets events can mimic the single top signal when the Z decays leptonically to e^+e^- or $\mu^+\mu^-$ at the same time as one of these leptons is not reconstructed. This background is significantly smaller than W+jets.

Dibosons: WW and WZ each has a similar signature as single top when one W decays to $\ell\nu$ while the other boson decays to quarks. ZZ events might mimic our signal when one Z decays to jets, and the other to leptons of which one is not reconstructed. However, the cross sections for these processes are small, and diboson events are hence not a major background for single top.

5.4 Signal and Background Modeling

5.4.1 Monte Carlo Simulation

The signal samples and the W+jets, $t\bar{t}$, Z+jets and diboson samples are generated using Monte Carlo simulations. In all cases, PYTHIA version 6.409 [1] with the Tune A settings is used to simulate the underlying event, initial and final state radiation, and the hadronization. For all background samples, the flavour and momentum of each participating parton inside the proton or antiproton are modelled by the CTEQ6L1 set of parton density functions [2], signal used CTEQ6M.

All stable particles produced are passed through a full detector simulation that models the interactions between the particles and the material in the detector using GEANT [3]. The magnetic field is also simulated such that charged particle trajectories are bent as they travel through the detector.

The electronic response due to the deposited energy is modeled by a program called DØSIM [4], which also simulates electronic noise and adds detector signals from zero-bias events to account for additional hard-interactions. Zero-bias events are data events recorded with no trigger requirements.

The final step in the Monte Carlo generation process is to reconstruct the event in the same way as a real data event is reconstructed (see Chap. 4). Due to the detector upgrade, but also due to changes in the software framework, it is necessary to create separate samples corresponding to the Run IIa and Run IIb run periods.

5.4.2 Monte Carlo Signal Samples

The single top quark events used in this analysis are generated by the Monte Carlo event generator SINGLETOP [5] which is based on the COMPHEP generator. The top mass is set to 170 GeV for the simulations.

The s-channel MC is produced using the leading order matrix element, and is scaled by an NLO/LO k-factor. The resulting kinematic distributions for all partons agree with the expectations from NLO calculations [6].

The situation is more complicated for the t-channel, where the higher order diagram $gq \rightarrow tq'b$ has an effective cross section on the same order as the LO diagram $bq \rightarrow tq'$ (Feynman diagram for these processes are shown in Fig. 2.3). These modes need to be combined to properly model the NLO kinematics. In order

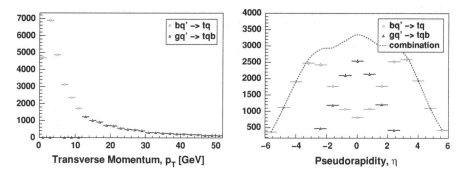

Fig. 5.2 The single top t-channel Monte Carlo simulation is generated as a mixture of the $2 \rightarrow 2$ and $2 \rightarrow 3$ modes. The *left plot* shows the p_T distributions for the b quark produced in association with the top. The matching of the modes can be seen at 12 GeV. The *right plot* shows corresponding pseudo-rapidity distributions and their sum

to avoid overlap, it is necessary to add requirements to the p_T of the b quark produced in association with the top. In case of the $gq \rightarrow tq'b$, this quark is added by PYTHIA as an ISR b quark produced from gluon splitting. SINGLETOP generates $bq \rightarrow tq'$ events with the restriction $p_T(\bar{b}) < 12$ GeV for the quark added by PYTHIA, and $gq \rightarrow tq'\bar{b}$ events requiring $p_T(\bar{b}) > 12$ GeV. The modes are generated in proportions such that the $p_T(\bar{b})$ spectrum becomes smooth as is shown in Fig. 5.2.

5.4.3 Monte Carlo Background Samples

The W+jets, $t\bar{t}$ and Z+jets backgrounds are modelled by ALPGEN [7] version 2.11, which is a leading order matrix element event generator. Separate samples are generated with different number of final state partons in order to properly simulate events with high jet multiplicities. The MLM matching scheme [8], which is provided within ALPGEN, is applied after the parton showering process in order to avoid overlap between the subsamples with different parton multiplicities.

The MLM scheme works as follows: Parton jets are reconstructed using the UA1 jet algorithm [9] with cone size 0.4. This is done after PYTHIA has applied parton showering and initial and final state radiation. Each tree-level parton generated by ALPGEN is required to match a jet with transverse momentum greater than 8 GeV within $\Delta R < 0.7$. If all tree-level partons fulfil these matching criteria, then the inclusive MLM matching criterion is met. If all partons are matched, and there are no additional unmatched parton jets in the event, then the exclusive MLM matching criterion is satisfied. Inclusive matching hence allows extra jets to be created by PYTHIA during parton showering. This matching is only used for the subsample with the highest parton multiplicity (see Tables 5.2 and 5.3).

The W+jets events have a leptonically decaying W boson and 0–5 partons in the final state. The factorization scale used is $m_W^2 + \sum m_T^2$, where m_T is the transverse

Table 5.2 The ALPGEN leading log cross sections provided during generation, the MLM matching applied, an approximate NLO/LL k-factor, and the number of generated Run IIa and Run IIb events

Process	Matching	$\mathcal{B}\sigma$ [pb]	k-factor	Run IIa stat.	Run IIb stat.
W+jets Monte Carlo sample details					
$W + 0lp \rightarrow \ell v + 0lp$	excl.	4550	1.30	2.9M	10.1M
$W + 1lp \rightarrow \ell v + 1lp$	excl.	1277	1.30	8.5M	3.5M
$W + 2lp \rightarrow \ell v + 2lp$	excl.	299	1.30	5.0M	2.3M
$W + 3lp \rightarrow \ell v + 3lp$	excl.	70.1	1.30	2.4M	1.1M
$W + 4lp \rightarrow \ell v + 4lp$	excl.	15.8	1.30	1.7M	1.0M
$W + 5lp \rightarrow \ell v + 5lp$	incl.	5.27	1.30	0.5M	0.2M
Wjj total		6217	1.30	21.0M	18.3M
$Wb\bar{b} + 0lp \rightarrow \ell vb\bar{b} + 0lp$	excl.	9.34	1.91	1.2M	1.4M
$Wb\bar{b} + 1lp \rightarrow \ell vb\bar{b} + 1lp$	excl.	4.27	1.91	0.6M	1.0M
$Wb\bar{b} + 2lp \rightarrow \ell vb\bar{b} + 2lp$	excl.	1.55	1.91	0.2M	0.6M
$Wb\bar{b} + 3lp \rightarrow \ell vb\bar{b} + 3lp$	incl.	0.74	1.91	0.2M	0.4M
$Wb\bar{b}$ total		15.9	1.91	2.3M	2.5M
$Wc\bar{c} + 0lp \rightarrow \ell vc\bar{c} + 0lp$	excl.	24.0	1.91	1.0M	1.0M
$Wc\bar{c} + 1lp \rightarrow \ell vc\bar{c} + 1lp$	excl.	13.4	1.91	0.6M	0.9M
$Wc\bar{c} + 2lp \rightarrow \ell vc\bar{c} + 2lp$	excl.	5.38	1.91	0.3M	0.5M
$Wc\bar{c} + 3lp \rightarrow \ell vc\bar{c} + 3lp$	incl.	2.51	1.91	0.3M	0.5M
$Wc\bar{c}$ total		45.3	1.91	2.3M	3.0M

Table 5.3 Information of the $t\bar{t}$ samples

Process	Matching	$\mathcal{B}\sigma$ [pb]	k-factor	Run IIa stat.	Run IIb stat.
Top pair Monte Carlo sample details					
$t\bar{t} + 0lp \rightarrow \ell vbb + 0lp$	excl.	1.51	1.42	1.4M	0.7M
$t\bar{t} + 1lp \rightarrow \ell vbb + 1lp$	excl.	0.62	1.42	0.8M	0.4M
$t\bar{t} + 2lp \rightarrow \ell vbb + 2lp$	incl.	0.31	1.42	0.4M	0.2M
Total $t\bar{t} \rightarrow \ell +$ jets		2.44	1.42	2.6M	1.3M
$t\bar{t} + 0lp \rightarrow \ell\ell vvbb + 0lp$	excl.	0.38	1.36	0.7M	0.3M
$t\bar{t} + 1lp \rightarrow \ell\ell vvbb + 1lp$	excl.	0.16	1.36	0.4M	0.6M
$t\bar{t} + 22p \rightarrow \ell\ell vvbb + 2lp$	incl.	0.08	1.36	0.2M	0.1M
Total $t\bar{t} \rightarrow \ell\ell +$ jets		0.61	1.36	1.3M	1.0M

The MLM matching applied, the ALPGEN leading log cross sections, the NLO k-factor applied, and the number of Run IIa and Run IIb events generated. The k-factor is calculated by dividing the theoretical NLO cross section for $t\bar{t}$ production (see Table 5.4) with the alpgen cross section

mass defined as $m_T^2 = m^2 + p_T^2$ and the sum $\sum m_T^2$ extends over all final state partons. Separate subsamples are generated as described below in order to ensure good statistics for the important W+heavy flavour events and to properly model events with many jets.

Wlp: These samples are created from diagrams with the final states $W + Nlp \rightarrow \ell v + Nlp$, where $N \in \{0, 1, 2, 3, 4, 5\}$, and lp is short for "light parton", meaning a gluon or a massless u, d, s or c quark. The sample is further divided into the subsets Wcj, meaning $Wc + N'lp \rightarrow \ell\nu c + N'lp(N' = N - 1)$, and Wjj, meaning processes without any final state c quarks.

Wbb: It denotes $Wb\bar{b} + Nlp \rightarrow \ell\nu b\bar{b} + Nlp$, where the two b quarks are massive, and $N \in \{0, 1, 2, 3\}$.

Wcc: It denotes $Wc\bar{c} + Nlp \rightarrow \ell\nu c\bar{c} + Nlp$. The c quarks are massive, and $N \in \{0, 1, 2, 3\}$. Events with two c quarks after parton showering are removed from the Wlp and Wbb samples, as well as events with b quarks in the Wlp sample, such that there is no phase-space overlap between the samples [10]. Further details about the W+jets subsamples are given in Tables 5.2 and 5.4.

The Z+jets samples are generated similarly to the W+jets samples. The Z bosons are set to decay leptonically, and the factorization scale used is $m_Z^2 + \sum m_T^2$. Separate samples for the Zjj, Zbb and Zcc processes are generated with up to four partons in the final state. Details about these samples can be seen in Table 5.4.

The $t\bar{t}$ samples either have one of the W bosons decaying to $\ell\nu$ while the other decays to two quarks (ℓ+jets), or both W bosons decay leptonically (dilepton). Matrix elements for $t\bar{t}$ production with 0–2 additional light partons are used. The top quark mass is set to 170 GeV (just as for the signal sample), and the

Table 5.4 The cross sections, branching fractions, and initial numbers of events in the Monte Carlo event samples

Event type	Cross section (pb)	Branching fraction	Run IIa statistics	Run IIb statistics
Monte Carlo sample overview				
Signals				
$tb \rightarrow \ell$+jets	$1.12^{+0.05}_{-0.12}$	0.3240 ± 0.0032	0.6M	0.8M
$tqb \rightarrow \ell$+jets	$2.34^{+0.13}_{-0.17}$	0.3240 ± 0.0032	0.5M	0.8M
Signal total	$3.46^{+0.18}_{-0.29}$	0.3240 ± 0.0032	1.1M	1.6M
Backgrounds				
$t\bar{t} \rightarrow \ell$+jets	$7.91^{+0.61}_{-1.01}$	0.4380 ± 0.0044	2.6M	1.3M
$t\bar{t} \rightarrow \ell\ell$	$7.91^{+0.61}_{-1.01}$	0.1050 ± 0.0010	1.3M	0.9M
Top pairs total	$7.91^{+0.61}_{-1.01}$	0.5430 ± 0.0054	3.9M	2.2M
$Wb\bar{b} \rightarrow \ell\nu bb$	93.8	0.3240 ± 0.0032	2.3M	2.5M
$Wc\bar{c} \rightarrow \ell\nu cc$	266	0.3240 ± 0.0032	2.3M	3.0M
$Wjj \rightarrow \ell\nu jj$	24,844	0.3240 ± 0.0032	21.0M	18.3M
W+jets total	25,205	0.3240 ± 0.0032	25.6M	23.8M
$Zb\bar{b} \rightarrow \ell\ell bb$	43.0	0.10098 ± 0.00006	1.0M	1.0M
$Zc\bar{c} \rightarrow \ell\ell cc$	114	0.10098 ± 0.00006	0.2M	1.0M
$Zjj \rightarrow \ell\ell jj$	7,466	0.10098 ± 0.00006	3.9M	7.0M
Z+jets total	7,624	0.03366 ± 0.00002	5.1M	9.0M
$WW \rightarrow$ anything	12.0 ± 0.7	1.0 ± 0.0	2.9M	0.7M
$WZ \rightarrow$ anything	3.68 ± 0.25	1.0 ± 0.0	0.9M	0.6M
$ZZ \rightarrow$ anything	1.42 ± 0.08	1.0 ± 0.0	0.9M	0.5M
Diboson total	17.1 ± 1.0	1.0 ± 0.0	4.7M	1.8M

The symbol ℓ stands for lepton (electron, muon or tau)

factorization scale to $m_t^2 + \sum p_T^2(\text{jets})$. Details for these samples are given in Tables 5.3 and 5.4.

Samples for the diboson processes WW, WZ, and ZZ are generated using PYTHIA. There are no constraints on the decays of the bosons. Some details about these samples are presented in Table 5.4.

5.4.4 Monte Carlo Corrections

The Monte Carlo simulations described in Sect. 5.4.1 model the particle interactions and the detector response. However, aspects of wear-and-tear of the detector are not considered, for example debris build-up and ageing effects. As a result, reconstruction efficiencies for electrons, muons and jets tend to be overestimated in the simulations. The energy and momentum resolutions for jets and leptons are also better in the simulated samples relative to data.

To account for these effects, scale factors and smearing factors are applied to the Monte Carlo Samples. The smearing factors used in this analysis are random shifts sampled from a Gaussian distribution. These factors are used to adjust the reconstructed energies and momenta of the simulated objects such that the resolution in Monte Carlo agrees with the resolution in data.

The following subsections describe the corrections that are applied to the simulated samples in order to reach agreement with data.

5.4.4.1 Primary Vertex Position

The distribution of the z position of the primary interaction point tends to be wider in data than it is in the simulation. A correction factor (weight) is applied to each simulated event depending on the z position of the primary vertex, the data epoch (Run IIa or Run IIb) and the instantaneous luminosity [11, 29]. The weight applied is about 1.5 for events with large $|z|$ (≈ 50 cm) and close to unity for events with a central primary vertex.

5.4.4.2 Instantaneous Luminosity Reweighting

The instantaneous luminosity for a simulated event is determined from the corresponding value for the overlayed zero-bias data event (see Sect. 5.4.1). The instantaneous luminosity is proportional to the average number of additional $p\bar{p}$ interactions and since the vast majority of additional collisions result in dijet events, the instantaneous luminosity is also correlated with the number of additional jets. The simulation does not do a perfect job when picking the overlay events. A weight depending on the instantaneous luminosity and the data epoch is assigned to each simulated event such that the luminosity spectrum for each individual Monte Carlo sample agrees with the spectrum observed in data.

5.4.4.3 Zp_T Reweighting

In the Z+jets samples, the Z p_T spectrum generated by ALPGEN does not quite agree with the next-to-leading order theory prediction. To account for this, a weight depending on the true Z p_T and the jet multiplicity is assigned to the event [12].

5.4.4.4 Electron Identification Efficiencies

Each event with an isolated electron is scaled by a factor that accounts for the differences in electron cluster finding and identification efficiency between data and Monte Carlo. The scale factor is divided into two parts: preselection and post–preselection. Preselection refers to the basic requirement for electron identification: the presence of an electromagnetic calorimeter cluster with a loose track match, electromagnetic fraction, and isolation. The preselection scale factor is parametrized in η_{det}. The post–preselection criteria consist of requirements on the H-matrix variable, track-matching and the likelihood. The post–preselection scale factor is parametrized in $(\eta_{\text{det}}, \phi)$. These factors are derived using $Z \to ee$ data and simulated events [13, 14].

The correction factor is given by:

$$\varepsilon_{e-\text{ID}} = \frac{\varepsilon_{\text{Presel}}^{\text{Data}}}{\varepsilon_{\text{Presel}}^{\text{MC}}} \times \frac{\varepsilon_{\text{PostPresel}}^{\text{Data}}}{\varepsilon_{\text{PostPresel}}^{\text{MC}}}.$$

5.4.4.5 Muon Efficiency Correction

The muon momenta in the Monte Carlo samples are smeared to match the resolution observed in data [15]. The muon smearing is parametrized in q/p_T and is determined in $Z \to \mu\mu$ events.

After the smearing is applied, a muon efficiency correction factor is calculated from three independent factors for identification, track matching and isolation efficiencies, according to

$$\varepsilon_{\mu-\text{ID}} = \frac{\varepsilon_{\text{Reco}}^{\text{Data}}}{\varepsilon_{\text{Reco}}^{\text{MC}}} \times \frac{\varepsilon_{\text{Track}|\text{Reco}}^{\text{Data}}}{\varepsilon_{\text{Track}|\text{Reco}}^{\text{MC}}} \times \frac{\varepsilon_{\text{Isolation}|\text{Track}}^{\text{Data}}}{\varepsilon_{\text{Isolation}|\text{Track}}^{\text{MC}}}.$$

This factor is applied to the event weight. The identification efficiency scale factor is parametrized in $(\eta_{\text{det}}, \phi)$, the track match scale factor is parametrized in track-z and η, and the isolation one in η.

5.4.4.6 Jet Corrections

Simulated jets have a better energy resolution, a higher reconstruction efficiency, and sometimes a higher average jet energy than what is observed in data. To correct for this, a procedure called JSSR (Jet Smearing Shifting and Removal) is applied at DØ [16]. The smearing and shifting parameters are measured as functions of jet p_T and η_{det} in direct photon events (γ+jets). The JSSR procedure only applies to jets with $p_T > 15$ GeV.

5.4.4.7 b Jet Identification Corrections

There are large differences for the track reconstruction efficiency between simulated samples and data. The tracking efficiency is significantly higher in Monte Carlo. One cannot directly apply the data neural network b-tagger to the simulated events since the algorithm relies heavily on tracks. Instead the probability to tag a b jet, a charm jet or a light jet is measured in data and applied to the Monte Carlo events. These probabilities are called Tag Rate Functions (TRFs).

In order to apply the b-tagging algorithm to a jet, it has to be *taggable*, meaning that there has to be a set of tracks associated with the jet. The probability for a jet to be taggable is also higher in Monte Carlo than in data, so an additional taggability correction must be applied. The probability P_{tag} for a jet to be b-tagged can be written as

$$P_{\text{tag}}(p_T, \eta, z, f) = \epsilon_{\text{taggable}}(p_T, \eta, z_{vtx}, f)\text{TRF}(p_T, \eta, z_{vtx}, f), \qquad (5.1)$$

where η and p_T are for the jet, f is the flavour of the jet: b, c or light, $\epsilon_{\text{taggable}}$ is the taggability efficiency for jets in data, and TRF is the tag rate function also for jets in data. z_{vtx} is the z-position of the vertex associated with the jet.

To simulate the b tagging in the Monte Carlo samples, several *permutations* of each event are created where each jet is set to either be b-tagged or not. If there are N_{jets} jets in an event, then $2^{N_{\text{jets}}}$ such permutations can be created. For instance, four permutations can be created when an event has two jets: both jets can be tagged, both jets can be untagged, or either of the two jets can be tagged while the other is not. The probability for a given permutation to occur is given by

$$\prod_{i}^{N_{\text{jets}}} \left(I_{\text{tag}}(i)P_{\text{tag}}(i) + (1 - I_{\text{tag}}(i))(1 - P_{\text{tag}}(i)) \right), \qquad (5.2)$$

where P_{tag} is given by Eq. 5.1, and $I_{\text{tag}}(i)$ is 1 if jet i is b-tagged and 0 if it is not. The probabilities for all permutations add up to unity.

For each simulated event, all possible b-tagging permutations are created, and each permutation is weighted by its probability according to Eq. 5.2. All the permutations, except the ones with zero probability, are considered for event selection.

As described in Sect. 4.7, this analysis uses two different b-tagging operating points: LOOSE and TIGHT, meaning b-tag NN > 0.5 and NN > 0.775

respectively. More specifically, each event is required to have either exactly one jet satisfying TIGHT b-tagging while the other jets do not satisfy LOOSE b-tagging, or exactly two jets satisfying LOOSE b-tagging. Separate tag rate functions are derived for the TIGHT and LOOSE operating points. The permutation weight for jet i being a TIGHT b jet while all other jets are not LOOSE can be written as:

$$P_{\text{tag}}^{\text{TIGHT}}(i) \prod_{\substack{j \neq i}}^{N_{\text{jets}}} (1 - P_{\text{tag}}^{\text{LOOSE}}(j)). \tag{5.3}$$

The permutation weights for two jets fulfilling LOOSE b-tagging can be calculated using the general formula (Eq. 5.2) with P_{tag} set to $P_{\text{tag}}^{\text{LOOSE}}$. For example, the permutation of an event with three jets where jet 1 and 3 are b tagged ($I_{\text{tag}}(1) = I_{\text{tag}}(3) = 1$, $I_{\text{tag}}(2) = 0$) will get the permutation weight:

$$P_{\text{tag}}^{\text{LOOSE}}(1)(1 - P_{\text{tag}}^{\text{LOOSE}}(2))P_{\text{tag}}^{\text{LOOSE}}(3).$$

5.4.4.8 W+jets Reweighting

After comparing with data, it is found that ALPGEN mismodels some of the kinematic variables, in particular the number of forward jets. To deal with this, the W+jets samples are reweighted before b-tagging to reach agreement with the jet η distributions observed in data. The reweighting is derived by comparing the W+jets sample to the data after subtraction of all other backgrounds. Reweighting functions are derived for the following variables in the order specified: leading jet η, second leading jet η, the $\Delta\phi$ and $\Delta\eta$ between the two leading jets, and thereafter the third and fourth jet η when applicable.

These reweighting functions are derived such that the overall normalization stays the same. Only the kinematic shape of the sample is affected.

5.4.5 Monte Carlo Sample Normalization

The $t\bar{t}, Z$+jets, dibosons, and single top samples are normalized to the integrated luminosity (Eq. 3.3) of the dataset using the cross sections and branching fractions listed in Table 5.4. Thereafter the corrections described in Sect. 5.4.4 are applied, and no further normalization is necessary.

The W+jets background is corrected in the same way as the other Monte Carlo samples, but here further corrections are needed. The sample is first normalized to the ALPGEN leading log cross sections listed in Table 5.2, but these cross sections have large uncertainties and are very sensitive to renormalization and factorization scale choices. Also, the higher order corrections to the cross section calculations are quite large, and from comparisons with NLO calculations, it is clear that the

Table 5.5 Scale factors applied to the W+jets sample

Subset	Wjj	Wcj	Wbb	Wcc
W+jets Scale Factors				
k-factor	1.3	1.8	1.91	1.91
S_{HF}	–	–	0.95	0.95

The k-factor is correcting the ALPGEN leading log cross section normalization to NLO, and the S_{HF} factor is measured by comparing the simulated samples to data [17]. The final scale factors needed to reach agreement with data are listed in Table 5.6 (see Sect. 5.4.6)

amount of W+jets is underpredicted. Approximate NLO/LL k factors are listed in Table 5.5. These k-factors are applied, but from comparison with data, it is clear that further scaling of the W+jets is needed.

The final W+jets normalization factors are derived from comparison with data. The W heavy flavour components Wbb and Wcc are adjusted by the scale factor $S_{HF} = 0.95 \pm 0.13$, which is calculated from the b tagging efficiencies in the subset of the data that contains two jets [17]. This subset is dominated by W+jets since $t\bar{t}$ events tend to have more jets. The final normalization factors applied to W+jets are the described in Sect. 5.4.6.

5.4.6 Multijets and W+jets Normalization

Multijet events enter the dataset by faking an isolated lepton and $\not{E}_T$. To model this background, a data sample is created using the same selection criteria as for the main analysis (Sect. 5.5), but an "inverted" lepton identification criterion. For the electron channel, the reconstructed electrons are no longer required to have a track match, and the likelihood cut is inverted: $\mathcal{L}_{EM} < 0.85$. For the muon channel, the muon isolation criterion is dropped, and events with a muon fulfilling tight muon isolation are rejected.

The data sample resulting from this selection is orthogonal to the analysis dataset since no events satisfy the tight lepton requirements. The reconstructed lepton is highly probable to be a fake lepton since the lepton identification criteria are very loose at the same time as tight leptons are rejected.

Two scale factors, $S_{W+\text{jets}}$ and $S_{\text{multijets}}$, are applied to the W+jets and multijet samples respectively. They are derived such that the total number of predicted events match data before any b tagging selection is applied. These scale factors hence fulfill the relation:

$$N_{\text{data}} = S_{W+\text{jets}} Y_{W+\text{jets}} + S_{\text{multijets}} Y_{\text{multijets}} + \mathcal{Y}_{\text{all other MC}}, \qquad (5.4)$$

where N_{data} are the number of events in data, $Y_{W+\text{jets}}$ and $Y_{\text{multijets}}$ are the sum of weights for all events in the W+jets and multijet samples, and $\mathcal{Y}_{\text{all other MC}}$ are the predicted number of events for the remaining signal and background samples normalized as described in Sect. 5.4.5. Notice, that since all terms but $S_{W+\text{jets}}$ and $S_{\text{multijets}}$ are known in Eq. 5.4, these factors are anticorrelated, and we only have one unknown parameter.

Table 5.6 W+jets and multijets normalization scale factors derived as described in Sect. 5.4.6

	$S_{W+\text{jets}}$				$S_{\text{multijets}}$			
	Run IIa		Run IIb		Run IIa		Run IIb	
	e	μ	e	μ	e	μ	e	μ
W+jets and Multijet KS Scale Factors								
2 jets	1.51	1.30	1.41	1.23	0.348	0.0490	0.388	0.0639
3 jets	1.92	1.79	1.75	1.57	0.291	0.0291	0.308	0.0410
4 jets	2.29	2.06	1.81	1.92	0.189	0.0244	0.424	0.0333

The $S_{W+\text{jets}}$ and $S_{\text{multijets}}$ are determined by comparing the lepton p_T, $\not{E}_T$, and $m_T(W)$ distributions between data and background, which all have significantly different shapes in W+jets and multijet backgrounds. The calculation of the W transverse mass is described in Sect. 7.1.2.

The procedure according to which $S_{W+\text{jets}}$ and $S_{\text{multijets}}$ are calculated is the following:

1. Set $S_{W+\text{jets}} = 1.0$ and calculate the corresponding $S_{\text{multijets}}$ from Eq. 5.4
2. Do a Kolmogorov–Smirnov test (KS-test) between data and background for the each of the p_T, $\not{E}_T$, and $m_T(W)$ distributions and record the KS-test values
3. Increase $S_{W+\text{jets}}$ by 0.001
4. Repeat from step 2 until $S_{W+\text{jets}}$ reaches 4.0 or when $S_{\text{multijets}}$ becomes negative
5. For each of the three variables, select the recorded $(S_{W+\text{jets}}, S_{\text{multijets}})$ which gave the highest KS-test value
6. The final scale factors are the weighted average of the three scale factors selected in step 5, using the KS-test value as weight.

The procedure above is done individually for electrons and muons and each jet multiplicity bin. The derived scale factors are listed in Table 5.6.

5.5 Event Selection Criteria

The event selection is designed to find events with a leptonically decaying W boson and jets. Each event is required to have an isolated lepton, missing transverse energy from the neutrino and two to four jets. The selection is applied separately for the electron and muon data.

5.5.1 General Selection

- Good quality (for data)
- Trigger requirement: at least one of the selected triggers has to fire (see Sect. 5.2)
- Good primary vertex: $|z_{PV}| < 60$ cm with at least three tracks attached
- 2–4 good jets with $p_T > 15$ GeV and $|\eta^{\text{det}}| < 3.4$

- The leading jet is required to have $p_T > 25$ GeV
- Missing transverse energy

  $20 < \not{E}_T < 200$ GeV in events with exactly two good jets
  $25 < \not{E}_T < 200$ GeV in events with three or more good jets

5.5.2 b-Tagging Selection

- Each jet must have $|\eta| < 2.5$ to be considered for b-tagging
- One jet fulfilling the TIGHT b-tagging criterion (NN > 0.775) while the other jets do not fulfill LOOSE b-tagging (NN < 0.5), or two jets fulfilling LOOSE b-tagging
- The leading b-tagged jet is required to have $p_T > 20$ GeV

5.5.3 Electron Channel Selection

- One tight electron with $|\eta^{\mathrm{det}}| < 1.1$ and $p_T > 15$ (20) GeV in events with 2 (3 or more) good jets
- No additional loose electron with $p_T > 15$ GeV
- No tight isolated muon with $p_T > 15$ GeV and within $|\eta^{\mathrm{det}}| < 2.0$
- Electron track pointing back to the primary vertex: $|\Delta z(e, \mathrm{PV})| < 1$ cm

5.5.4 Muon Channel Selection

- One tight muon with $p_T > 15$ GeV and $|\eta^{\mathrm{det}}| < 2.0$
- No additional loose muons with $p_T > 4$ GeV
- No loose electron with $p_T > 15$ GeV and within $|\eta^{\mathrm{det}}| < 2.5$
- Muon track pointing back to the primary vertex: $|\Delta z(\mu, \mathrm{PV})| < 1$ cm
- Additional $p_T > 30$ GeV criterion applied to the leading jet when it is in the intercryostat region $1.0 < |\eta^{\mathrm{det}}| < 1.5$

The selection criteria listed up to this point select a significant amount of multijet background. It is desirable to reduce this background since it is difficult to model, especially when the $\not{E}_T$ is parallel or back-to-back with a (mis)reconstructed object.

The following selection criteria have been designed to reduce the amount of multijet background while keeping most of the signal:

5.5.5 Multijet Reduction Criteria

- Various angular selection criteria that remove events with low $\not{E}_T$ at the same time as the $\not{E}_T$ vector is either back-to-back or parallel to the lepton or the leading jet (see Figs. 5.3 and 5.4)

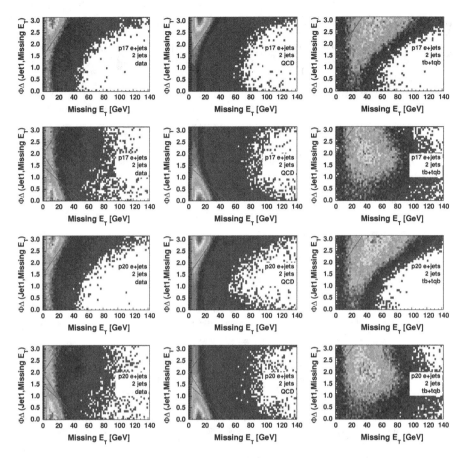

Fig. 5.3 $\Delta\phi$ (jet1, $\not{E}_T$) versus $\not{E}_T$ (*first* and *third rows*) and $\Delta\phi$ (lepton, $\not{E}_T$) versus $\not{E}_T$ (*second* and *fourth rows*) distributions for data (*left*), multijets (*centre*) and $tb + tqb$ signal (*right*), in the electron channels in Run IIa (*two first rows*) and Run IIb (*second two rows*) data. The "triangular" selection criteria applied are given by the lines in the plots. All events are required to fall to the right of the lines shown. The events that fail these cuts have low $\not{E}_T$ at the same time as the $\not{E}_T$ is aligned or anti-aligned with a reconstructed object in the event

- Selection on the scalar sum of the $\not{E}_T$ and the p_T of the lepton and all jets. In the electron channel:

  - $H_T(\text{alljets}, e, \not{E}_T) > 120$ GeV for events with $N_{\text{jets}} = 2$
  - $H_T(\text{alljets}, e, \not{E}_T) > 140$ GeV for events with $N_{\text{jets}} = 3$
  - $H_T(\text{alljets}, e, \not{E}_T) > 160$ GeV for events with $N_{\text{jets}} = 4$

  In the muon channel:

  - $H_T(\text{alljets}, \mu, \not{E}_T) > 110$ GeV for events with $N_{\text{jets}} = 2$
  - $H_T(\text{alljets}, \mu, \not{E}_T) > 130$ GeV for events with $N_{\text{jets}} = 3$
  - $H_T(\text{alljets}, \mu, \not{E}_T) > 160$ GeV for events with $N_{\text{jets}} = 4$

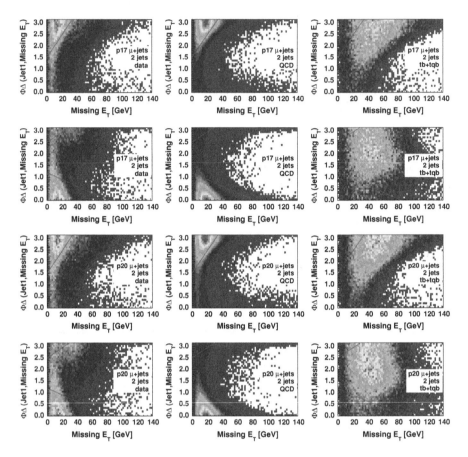

Fig. 5.4 $\Delta\phi$ (jet1, $\not{E}_T$) versus $\not{E}_T$ (*first* and *third rows*) and $\Delta\phi$ (lepton, $\not{E}_T$) versus $\not{E}_T$ (*second* and *fourth rows*) distributions for data (*left*), multijets (*centre*) and *tb+tqb* signal (*right*), in the muon channels in Run IIa (*two first rows*) and Run IIb (*second two rows*) data. The "triangular" selection criteria applied are given by the lines in the plots. All events are required to fall to the right of the lines shown. The events that fail these cuts have low $\not{E}_T$ at the same time as the $\not{E}_T$ is aligned or anti-aligned with a reconstructed object in the event

5.6 Event Yields

The number of events in data, and the predicted number of signal and background events are here referred to as *yields*. The yields after the selection described in Sect. 5.5 are presented for the Run IIa dataset in Table 5.7 and for Run IIb in Table 5.8.

The Run IIa data, signal and background yields for events with exactly one *b*-tagged jet, are presented in Table 5.9, and the corresponding yields for Run IIb are shown in Table 5.10. The yields for events with exactly two *b*-tagged jets are given in Tables 5.11 and 5.12 for the Run IIa and Run IIb datasets respectively.

Table 5.7 Yields for Run IIa data, signal and all background components after event selection

	Electron channel			Muon channel		
	2 jets	3 jets	4 jets	2 jets	3 jets	4 jets
Run IIa event yields before b tagging						
Signals						
tb	23	8.2	2.2	27	11	2.8
tqb	43	17	5.8	51	23	7.0
$tb+tqb$	67	26	8.0	77	33	10
Backgrounds						
$t\bar{t} \to \ell\ell$	60	37	12	57	41	13
$t\bar{t} \to \ell$+jets	41	136	158	32	143	196
$Wb\bar{b}$	479	160	45	530	211	57
$Wc\bar{c}$	1,041	356	101	1,196	485	125
Wcj	1,338	315	65	1,514	389	81
Wjj	13,847	3,309	722	17,028	4,612	984
$Zb\bar{b}$	18	7.1	3.5	70	22	6.6
$Zc\bar{c}$	33	12	4.2	151	46	13
Zjj	461	125	40	1,309	348	84
Dibosons	339	98	24	457	142	34
Multijets	923	278	74	896	235	69
Background Sum	18,582	4,834	1,246	23,243	6,675	1,663
Data	18,582	4,834	1,246	23,243	6,675	1,663

No b tagging requirements have been applied

Table 5.8 Yields for Run IIb data, signal and all background components after event selection

	Electron channel			Muon channel		
	2 jets	3 jets	4 jets	2 jets	3 jets	4 jets
Run IIb event yields before b tagging						
Signals						
tb	24	8.4	2.3	26	11	3.3
tqb	43	19	6.3	49	24	7.9
$tb+tqb$	67	27	8.6	75	35	11
Backgrounds						
$t\bar{t} \to \ell\ell$	65	42	13	61	46	14
$t\bar{t} \to \ell$+jets	43	141	168	33	145	198
$Wb\bar{b}$	458	161	42	499	200	61
$Wc\bar{c}$	1,006	351	94	1,126	453	137
Wcj	1,327	316	70	1,442	377	96
Wjj	14,166	3,489	795	16,941	4,710	1,137
$Zb\bar{b}$	19	8.2	3.4	70	26	7.4
$Zc\bar{c}$	35	15	5.9	152	54	14
Zjj	596	167	55	1,833	507	118
Dibosons	343	103	26	445	145	37
Multijets	987	294	188	1,369	377	108
Background sum	19,048	5,087	1,460	23,972	7,040	1,927
Data	19,048	5,087	1,460	23,972	7,040	1,927

No b tagging requirements have been applied

Table 5.9 Yields for Run IIa data, signal and all background components after event selection and requiring every event to have exactly one *b* tagged jet

	Electron channel			Muon channel		
	2 jets	3 jets	4 jets	2 jets	3 jets	4 jets
Run IIa single-tagged event yields						
Signals						
tb	9.1	3.1	0.82	10.2	3.9	1.0
tqb	17.4	6.6	2.1	20.5	8.6	2.6
tb+tqb	26.4	9.7	3.0	30.7	12.5	3.6
Backgrounds						
$t\bar{t} \to \ell\ell$	23.6	14.2	4.3	22.1	15.6	4.8
$t\bar{t} \to \ell$+jets	16.3	52.0	57.0	12.5	54.3	69.8
$Wb\bar{b}$	135.4	44.2	12.0	146.4	57.1	16.0
$Wc\bar{c}$	66.0	24.8	8.1	73.9	33.4	9.9
Wcj	98.3	24.0	5.0	112.1	30.2	6.3
Wjj	73.6	21.9	6.1	87.0	30.1	8.1
$Zb\bar{b}$	6.5	2.8	0.89	26.8	7.9	2.5
$Zc\bar{c}$	2.7	1.2	0.55	13.3	4.6	1.5
Zjj	5.4	1.8	0.63	12.7	4.3	1.1
Dibosons	16.2	5.3	1.4	22.3	7.8	2.1
Multijets	28.0	10.3	3.0	51.5	17.2	7.3
Background sum	472.1	202.4	99.0	580.6	262.5	129.4
Bkgds+signals	498.5	212.2	101.8	611.3	275.0	131
Data	508	202	103	627	259	131

Figure 5.5 illustrates the proportions of the signal and background components in the datasets classified by number of jets and number of *b*-tagged jets.

5.7 Data-Background Model Comparison

This section, and many of the subsequent sections in this thesis, show plots where the data are compared with the total background and signal predictions. Figure 5.6 shows the colour scheme used to label the data and the different signal and background components in these plots.

After all but the *b*-tagging selection criteria are applied, the dataset is referred to as the *pre-tag* sample (see Fig. 5.5). This sample is divided into 12 channels depending on the run period (Run IIa or Run IIb), the lepton flavour (*e* or *μ*), and the jet multiplicity (two, three or four jets). After *b*-tagging is applied, the total number of analysis channels grows to 24: 12 single-tagged and 12 double-tagged channels.

In order to ensure that the background is well modeled, the agreement between the data and the signal and background samples is studied for a long list of variables, both for each channel individually, and for various combinations of

Table 5.10 Yields for Run IIb data, signal and all background components after event selection and requiring every event to have exactly one b tagged jet

	Electron channel			Muon channel		
	2 jets	3 jets	4 jets	2 jets	3 jets	4 jets
Run IIb single-tagged event yields						
Signals						
tb	9.5	3.3	0.91	9.9	4.3	1.2
tqb	16.9	7.2	2.5	18.5	8.7	3.0
$tb+tqb$	26.4	10.5	3.4	28.4	13.0	4.2
Backgrounds						
$t\bar{t} \rightarrow \ell\ell$	25.6	16.1	4.9	23.5	17.0	5.0
$t\bar{t} \rightarrow \ell$+jets	16.4	53.8	61.5	12.2	53.6	70.6
$Wb\bar{b}$	129.5	44.3	11.6	136.0	53.4	16.8
$Wc\bar{c}$	68.6	26.2	7.6	72.4	32.6	10.9
Wcj	106.1	25.8	5.4	111.9	29.7	6.6
Wjj	114.4	35.4	9.6	128.3	46.5	14.5
$Zb\bar{b}$	5.0	2.4	1.0	20.1	7.7	2.2
$Zc\bar{c}$	2.1	1.1	0.57	10.7	4.3	1.2
Zjj	6.0	2.1	0.79	13.9	5.0	1.3
Dibosons	17.4	5.8	1.7	22.7	8.4	2.4
Multijets	31.0	10.1	7.1	73.5	28.2	9.0
Background sum	522.1	223.2	111.6	625.3	286.5	140.5
Bkgds+signals	548.5	233.6	115.2	653.6	299.4	144.7
Data	547	207	124	595	290	142

channels. This task is very time consuming due to the large number of variable and channel combinations. Several thousand plots are produced and checked. Initially, most attention is spent on the distribution of basic kinematic quantities. Examples of such distributions are shown in Fig. 5.7, where all 24 channels are combined. One of the variables shown in Fig. 5.7 is the W transverse mass defined by

$$m_T^2(W) = E_T^2(W) - \vec{p}_T^2(W) = (\not{E}_T + p_T(\ell))^2 - (\vec{\not{E}}_T + \vec{p}_T(\ell))^2. \quad (5.5)$$

This variable is expected to peak close to the mass of the W boson (around 80 GeV) for events containing real W bosons. The W transverse mass distribution for various combinations of channels, both before and after b-tagging, is shown in Fig. 5.8.

More details about the different variables and the agreement between data and the signal and background model is presented in Sect. 7.1.2.

5.8 Systematic Uncertainties

This section describes the different systematic uncertainties associated with the analysis. The relative uncertainties on each of the sources are summarized in Table 5.13, and presented in greater detail in Appendix B.

Table 5.11 Yields for Run IIa data, signal and all background components after event selection and requiring every event to have exactly two b tagged jets

	Electron channel			Muon channel		
	2 jets	3 jets	4 jets	2 jets	3 jets	4 jets
Run IIa double-tagged event yields						
Signals						
tb	5.69	2.12	0.58	6.66	2.75	0.75
tqb	0.81	1.61	0.90	0.92	2.15	1.07
$tb+tqb$	6.50	3.72	1.49	7.59	4.90	1.82
Backgrounds						
$t\bar{t} \rightarrow \ell\ell$	13.91	9.70	3.12	14.09	11.22	3.58
$t\bar{t} \rightarrow \ell$+jets	4.32	28.63	43.16	3.51	32.18	55.38
$Wb\bar{b}$	33.96	12.34	3.69	35.64	15.71	4.77
$Wc\bar{c}$	5.12	2.75	1.28	5.62	3.56	1.50
Wcj	1.44	0.68	0.19	1.62	0.83	0.26
Wjj	1.45	0.86	0.34	1.70	1.20	0.46
$Zb\bar{b}$	0.88	0.74	0.31	6.14	2.60	0.91
$Zc\bar{c}$	0.15	0.13	0.10	1.05	0.54	0.25
Zjj	0.14	0.09	0.05	0.31	0.20	0.07
Dibosons	2.05	0.85	0.28	3.06	1.37	0.46
Multijets	1.90	1.10	0.48	3.28	1.98	0.93
Background sum	65.33	57.88	53.00	76.03	71.40	68.57
Bkgds+signals	71.82	61.59	54.48	83.60	76.29	70.40
Data	67	61	37	71	62	56

- *Integrated luminosity* There is a 6.1% uncertainty on the integrated luminosity estimate, which affects the signal, $t\bar{t}$, Z+jets, and diboson yields.
- *Theory cross sections* For the single top and $t\bar{t}$ cross sections, there are uncertainties due to the scale, parton density functions, kinematics, and top quark mass choice that are combined in quadrature [11, 18]. The mass uncertainty is calculated as the difference between the cross section at 170 GeV (the value the analysis is performed at) and the most recent combined top mass measurement of 172.4 GeV [19]. The diboson cross section uncertainty is derived using the NLO MCFM generator [20]. The uncertainty for WW is 5.6%, for WZ 6.8%, and for ZZ 5.5%, and for the sum of the processes it is 5.8%. The average value of 5.8% is also used for the Z+jets background.
- *Branching fractions* The branching fractions for a W boson to decay to an electron, muon, or tau lepton, have an average uncertainty of 1.5% [21, 22]. This is one of the MC normalization uncertainties.
- *Parton distribution functions* The effect of changing the parton distribution functions is evaluated by reweighting each event in the single top Monte Carlo according to the 40 different CTEQ error PDFs. The systematic uncertainty affecting the signal acceptances from this source is estimated to be 3%.
- *Trigger efficiency* This analysis uses an OR of many trigger conditions which gives a trigger efficiency of close to 100%. The uncertainty of this trigger efficiency is measured to be 5% in all channels except for the Run IIb μ+jets

Table 5.12 Yields for Run IIb data, signal and all background components after event selection and requiring every event to have exactly two *b* tagged jets

	Electron channel			Muon channel		
	2 jets	3 jets	4 jets	2 jets	3 jets	4 jets
Run IIb double-tagged event yields						
Signals						
tb	5.26	2.00	0.58	5.61	2.59	0.78
tqb	0.94	1.89	1.01	0.99	2.22	1.21
tb+tqb	6.20	3.89	1.58	6.60	4.80	1.99
Backgrounds						
$t\bar{t} \to \ell\ell$	13.58	9.99	3.17	12.95	10.79	3.33
$t\bar{t} \to \ell$+jets	4.07	27.71	43.44	3.11	29.00	51.06
$Wb\bar{b}$	30.54	12.19	3.43	30.84	14.42	5.07
$Wc\bar{c}$	5.55	3.15	1.17	5.60	3.72	1.67
Wcj	2.04	0.96	0.28	2.07	1.04	0.33
Wjj	2.81	1.66	0.64	3.21	2.20	0.98
$Zb\bar{b}$	0.69	0.60	0.34	4.34	2.07	0.70
$Zc\bar{c}$	0.14	0.14	0.10	0.86	0.53	0.19
Zjj	0.16	0.11	0.06	0.34	0.24	0.09
Dibosons	1.96	0.91	0.30	2.98	1.38	0.46
Multijets	2.25	1.37	1.13	4.92	3.12	0.97
Background sum	63.78	58.80	54.06	71.22	68.50	64.85
Bkgds+signals	69.99	62.68	55.64	77.81	73.31	66.85
Data	79	56	51	85	79	80

Fig. 5.5 Illustration of the signal and background composition of the dataset depending on the number of jets and number of *b* tags

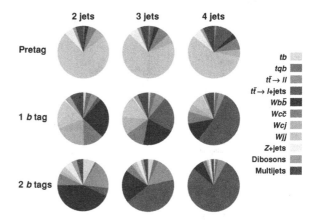

channels, where 10% is used. This uncertainty is treated as uncorrelated between Run IIa and Run IIb and between *e*+jets and μ+jets channels.

- *Instantaneous luminosity reweighting* The instantaneous luminosity distributions are reweighted for all Monte Carlo samples in order to match the Run IIa or Run IIb data distributions as appropriate. The initial distributions are from the zero-bias data overlaid on the MC events to simulate the underlying events, and

Fig. 5.6 The colour scheme
used to label signal and
background components (see
online version for plots in
colour). The order of the
components in the Key is the
same as the order in which
they are stacked in all data-
background comparison plots
shown in this thesis

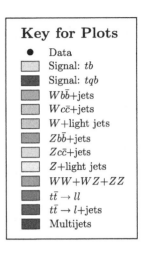

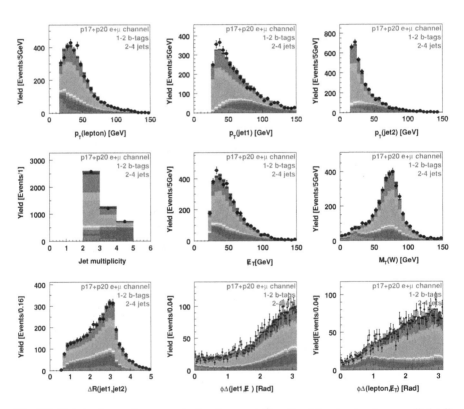

Fig. 5.7 Data-background agreement of various variables after b tagging has been applied (all
24 channels combined). A colour key for the signal and background components is shown in
Fig. 5.6

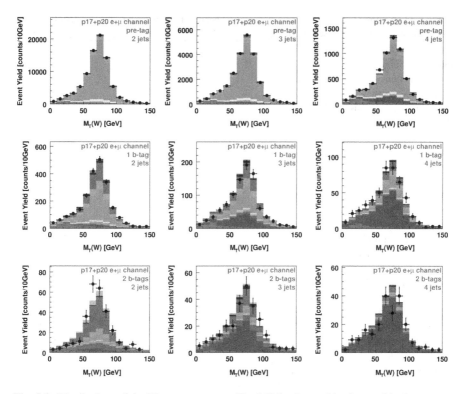

Fig. 5.8 Distributions of the W transverse mass (Eq. 5.5) for data and background in the pre-tag sample (*top row*), the single-tagged sample (*middle row*) and the double-tagged sample (*bottom row*). Channels with two, three and four jets are shown in the *left*, *middle* and *right* columns respectively. The colour key for the signal and background components can be seen in Fig. 5.6

are generally at too low values for later data-taking conditions. The uncertainty on this reweighting is 1.0%.

- *Primary vertex modeling and selection* The distribution of the z position of the primary vertex is reweighted in Monte Carlo to match that in data [11, 29]. The uncertainty due to this reweighting is 0.05% (negligible). The uncertainty on the difference in primary vertex selection efficiency between data and MC is 1.4%.
- *Electron reconstruction and identification efficiency* The electron scale factor uncertainty includes the dependence of the electron ID scale factor on the variables ignored in the parametrization, jet multiplicity dependence, track match and likelihood scale factor. The dependencies on ϕ and p_T of the electron are included in the systematic error as well and also the limited statistics in each bin of the parametrization. The assigned total uncertainty is 2.5%.
- *Muon reconstruction and identification efficiency* The MC scale factor uncertainties for muon reconstruction and identification, including isolation requirements, are estimated by the muon ID group as coming from the tag/probe

Table 5.13 A summary of the relative systematic uncertainties for each of the correction factors and normalizations scales used in the analysis

Components for normalization	
Integrated luminosity	6.1%
$t\bar{t}$ cross section	12.7%
Z+jets cross section	5.8%
Diboson cross sections	5.8%
Branching fractions	1.5%
Parton distribution functions (signal acceptances only)	3.0%
Triggers	5.0–10.0%
Instantaneous luminosity reweighting	1.0%
Primary vertex selection	1.4%
Lepton identification	2.5%
Jet fragmentation	0.7–4.0%
Initial- and final-state radiation	0.6–12.6%
b-jet fragmentation	2.0%
Jet reconstruction and identification	1.0%
Jet energy resolution	4.0%
W+jets heavy-flavour correction	13.7%
Z+jets heavy-flavour correction	13.7%
W+jets normalization to data	1.8–5.0%
Multijets normalization to data	30–54%
MC and multijets statistics	0.5–16%
Components for normalization and shape	
Jet energy scale for signal	1.1–13.1%
Jet energy scale for total background (not shape for Z+jets or dibosons)	0.1–2.1%
b tagging, single-tagged	2.1–7.0%
b tagging, double-tagged	9.0–11.4%
Component for shape only	
ALPGEN reweighting	–

method, background subtraction, and limited statistics in the parametrization. The assigned total uncertainty is 2.5%.

- *Jet fragmentation* The systematic uncertainty due to the modeling of the jet fragmentation is evaluated by comparing the acceptance of $t\bar{t}$ events generated with ALPGEN+PYTHIA (as used in the analysis) to ones generated with ALPGEN+HERWIG, with all other generation parameters unchanged. The resulting uncertainty is about 1–4%, and is applied to all MC samples in the analysis.
- *Initial-state and final-state radiation* This uncertainty is evaluated in $t\bar{t}$ samples generated with different amounts of initial- and final-state radiation. The uncertainty ranges from 0.6 to 12.6%.
- *b-jet fragmentation* The size of the uncertainty from the b-jet modeling is evaluated in the $t\bar{t}$ pairs cross section analysis following the method described in Ref. [26]. The uncertainty arises from the difference between the fragmentation parametrizations preferred by SLD vs. LEP data. A 2.0% value is measured.
- *Jet reconstruction and identification* The efficiency to reconstruct and identify jets is measured in both data and Monte Carlo [11, 12, 23, 24]. These

efficiencies are slightly higher for Monte Carlo, and a small correction is applied to the simulated samples. The uncertainty on the Monte Carlo normalization due to this correction is 1%.

- *Jet energy scale* The jet energy scale correction [25] is raised and lowered by one standard deviation on each MC sample and the whole analysis repeated, which produces a shape-changing uncertainty, and an overall normalization uncertainty. The normalization part ranges from 1.1 to 13.1% on the signal acceptance and from 0.1 to 2.1% on the combined background.

- *Jet energy resolution* A flat uncertainty of 4% is assigned due to the jet energy resolution. Using the method described in Ref. [16], it is found that the shape variations due to this uncertainty are smaller than 4% for all signals and backgrounds.

- ALPGEN*reweighting* The uncertainty due to the reweighting of the ALPGEN W+jets background affects the shapes of the W+jets background components (see Appendix B).

- *W +jets and multijets normalization* The W+jets and multijets background normalizations are determined from a fit to the pretagged data, as described in Sect. 5.4.6. The uncertainties from this fit vary from channel to channel and range from 30 to 54% for the multijets backgrounds and from 1.8 to 5.0% for the W+jets backgrounds.

- *Taggability and tag-rate functions for MC events* The uncertainty associated with b tagging in MC events is evaluated by adding the taggability [27] and the tag rate [28] components of the uncertainty in quadrature. The TRF uncertainties originate from several sources: statistical errors of the Monte Carlo event sets; the assumed fraction of heavy flavour in the multijets Monte Carlo events for the mistag rate determination; and, the TRF parametrizations. These uncertainties affect both shape and normalization of the Monte Carlo samples. The normalization part of the uncertainty is about 2.3% (9.9%) to 4.7% (10.8%) for single-tagged (double-tagged) signal acceptances, and from 2.1% (9.0%) to 7.0% (11.4%) for single-tagged (double-tagged) combined backgrounds. More details are given in Appendix B.

- *W+jets heavy-flavour scale factor correction* The heavy-flavour scale factor correction S_{HF} is measured in data [17]. The Monte Carlo tag rate function uncertainty induces fluctuations in the effective scale factor that are at least as large as the channel-to-channel variations in the measurement. Therefore, it can be argued that any additional systematic is double counting. However, an uncertainty of 13.7% is still assigned on the scale factor.

- *Z+jets heavy-flavour scale factor correction* The uncertainty used for the Z+heavy-flavour normalization scale factor is 13.7%, taken from the S_{HF} factor used for W+jets.

- *Sample statistics* The Monte Carlo and data samples used to estimate the signal and background shapes are limited in size. In particular, the number of multijets background events is quite low after b tagging. The statistical uncertainty on the different background components is taken into account for each sample in each bin of the final discriminant distribution.

References

1. T. Sjöstrand, S. Mrenna, P. Skands, PYTHIA 6.4 physics and manual, J. High Energy Phys. **0608**, 026 (2006)
2. J. Pumplin et al., New generation of parton distributions with uncertainties from global QCD analysis, J. High Energy Phys. **0207**, 012 (2002)
3. R. Brun, F. Carminati, GEANT: detector description and simulation tool, CERN Program Library Long Writeup, Report No. W5013 (1993)
4. DØsim, http://www-d0.fnal.gov/computing/MonteCarlo/simulation/d0sim.html
5. E.E. Boos et al., Method for simulating electroweak top-quark production events in the NLO approximation: singletop generator, Phys. Atom. Nucl. **69**, 1317 (2006)
6. Z. Sullivan, Understanding single-top-quark production and jets at hadron colliders, Phys. Rev. D **70**, 114012 (2004)
7. M.L. Mangano et al., ALPGEN, a generator for hard multiparton processes in hadronic collisions, J. High Energy Phys. **0307**, 001 (2003)
8. S. Höche, F. Krauss, N. Lavesson, L. Lönnblad, M.L. Mangano, A.Schälicke, S.: Schumann Matching parton showers and matrix elements. In: Roeck A., Jung H. (eds) Proceedings of HERA and the LHC: A Workshop on the Implications of HERA for LHCPhysics, p. 288. CERN, Geneva, (2005) . ISBN: 9290832657
9. R.K. Ellis, S. Veseli, Strong radiative corrections to $Wb\bar{b}$ production in $p\bar{p}$ collisions, Phys. Rev. D **60**, 011501 (1999)
10. D. Gillberg, Heavy flavour removal and determination of weighting factors for ALPGEN W+jets Monte Carlo, DØ Note 5129 (2006)
11. H. Schellman, The longitudinal shape of the luminous region at DØ, DØ Note 5142 (2006)
12. B. Tiller, T. Nunnemann, Measurement of the differential Z0-boson production cross-section as function of transverse momentum, DØ Note 4660 (2004)
13. J. Hays et al., Single electron efficiencies in p17 data and Monte-Carlo using p18.05.00 d0correct, DØ Note 5105 (2006)
14. O. Atramentov et al., Electron and photon identification with p20 data, DØ Note 5761 (2008)
15. T. Gadfort et al., Muon identification certification for p17 data, DØ Note 5157 (2006)
16. N. Makovec, J.-F. Grivaz, Shifting, smearing and removing simulated jets, DØ Note 4914 (2005)
17. Y. Peters et al., Study of the W+jets heavy flavor scale factor in p17, DØ Note 5406 (2007)
18. N. Kidonakis, Single Top Quark Production at the Fermilab Tevatron: Threshold Resummation and Finite-Order Soft Gluon Corrections, Phys. Rev. D **74**, 114012 (2006)
19. Tevatron Electroweak Working Group, CDF, D0 Collaborations,Combination of CDF and D0 Results on the Mass of the Top Quark,FERMILAB-TM-2466-E, arXiv:1007.3178v1 (2010)
20. The MCFM (N)NLO calculations of the diboson cross sections and their uncertainties are documented in http://www-clued0.fnal.gov/%7Enunne/cross-sections/mcfm_cross-sections.html
21. C. Amsler et al., Particle Data Group, Review of particle physics, Phys. Lett.B 667, 1 (2008).
22. B. Casey et al., Determination of the Run IIb luminosity constants, DØ Note 5559 (2007)
23. A.Harel, Jet ID Optimization, DØ Note 4919 (2006)
24. A.Harel, J.Kvita, p20 JetID Efficiencies and Scale Factors, DØ Note 4919 (2006)
25. A.Juste et al., DØ JES Group, Jet Energy Scale Determination at DØ Run 11, DØ Note 5382 (2007).
26. Y. Peters, M. Begel, K. Hamacher, D. Wicke, Reweighting of the fragmentation function for the DØ Monte Carlo, DØ Note 5325 (2007)
27. G. Otero, y. Garzón et al., Taggability in Pass2 p14 data, DØ Note 4995, (2006)
28. T. Gadfort et al., Performance of the DØ NN b-tagging tool on p20 data, DØ Note 5554, (2007)
29. H. Schellman, Run IIb longitudinal beam shape, DØ Note 5540 (2007)

Chapter 6
Analysis: Decision Trees

A *decision tree* is a multivariate technique which can be used to classify observations [1, 2]. In this thesis, the term decision trees refers to what is more specifically known as *classification trees*, and this technique is applied to separate single top quark events from a vast amount of background.

This chapter gives an overview of decision trees and motivates why and how they can be used in experimental particle physics.

6.1 Motivation

Single top production is a very rare process. After applying the event selection described in Chap. 5, the signal to background ratio is 1:20, and the signal excess is smaller than the uncertainty on the background prediction. In this situation, it is not possible to conduct a cross section measurement—better separation of signal from background is needed.

The traditional approach is to apply further selection criteria (cuts) on discriminating variables and select a subset of the original sample with an enhanced signal to background ratio. The main disadvantage with this method is that we lose precious signal every time a cut is applied.

A more effective way is to use a multivariate technique, where the separation power of several variables $\vec{x}$ is combined into a discriminant $D(\vec{x})$. This discriminant will separate signal from background better than any individual variable. The signal significance and cross section can then be calculated either by applying a selection criterion on $D(\vec{x})$, or, more effectively, by integrating over the full discriminant distribution.

The analysis described in this thesis uses boosted decision trees as a multivariate technique to derive a discriminant $D(\vec{x})$ that monotonically increases with the probability of an event being signal. The cross section and the signal significances are derived from the $D(\vec{x})$ distributions observed in data and expected for the signal and background processes using Bayesian calculations as described in Sect. 7.2.1.

D. Gillberg, *Discovery of Single Top Quark Production*, Springer Theses,
DOI: 10.1007/978-1-4419-7799-1_6, © Springer Science+Business Media, LLC 2011

6.2 Overview of Decision Trees

6.2.1 History and Usage

Decision trees originated in the fields of data mining and pattern recognition. Much of the initial development was done by Breiman et al. who developed the CART algorithm (Classification And Regression Trees) [1] in the early 1980s. Extensive studies of decision trees have been conducted since then resulting in a long list of publications mainly in different branches of computer science. Several methods that improve the classification performance by creating an ensemble (forest) of decision trees were developed in the 1990s (see Sect. 6.5). One of these extensions is *boosting* (Sect. 6.5.3) which is used in this analysis.

There are vast applications of decision trees in various fields including medical diagnostics, mass spectrum classification, financial analysis and hand writing recognition. In high energy physics, decision trees have rarely been used until quite recently. The two main applications are particle identification (PID) and isolation of a specific physics process from background processes (as in this analysis). Examples of PID applications include distinguishing jets originating from either a b quark or from the hadronic decay of a τ lepton from ordinary QCD jets. Boosted decision trees were first used in high energy physics by the MiniBooNE experiment [5, 6] for particle identification, and later by our group at Simon Fraser University as part of DØ's search for single top quark production [3, 4].

6.2.2 What is a Decision Tree?

A *binary tree*, or 2-tree, is a structure of *nodes* where each node can have up to two daughter nodes. The initial node is referred to as the *root node* and is typically assigned the identifier number $t = 1$. Left and right daughter nodes are assigned IDs $2t$ and $2t + 1$, respectively. The nodes are either *internal* (have daughter nodes) or *terminal* (no daughters). Terminal nodes are called *leaves*. An example of a binary tree is illustrated in Fig. 6.1.

A decision tree is a *n*-tree (up to *n* children per node) which can be used to classify observations into *n*-classes. Hereafter we will assume that we are dealing with only two classes, signal S and background B, in which case the decision tree is a binary tree. Internal nodes each have an associated test that, given the features of an observation $\vec{x}$, returns either true or false ("go right" or "go left"). Each leaf has an assigned decision tree output value.

An observation defined by variables $\vec{x}$ will, starting from the root node, follow a unique path through the decision tree depending on the outcomes of the tests from the internal nodes passed. Eventually the observation will end up at a leaf and the classification of the observation is the decision tree output value of this leaf. A simple decision tree is illustrated in Fig. 6.2.

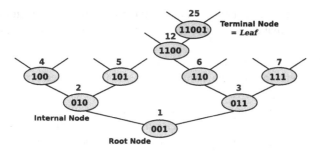

Fig. 6.1 Illustration of a binary tree. Each node is shown with its identifier number *t*

Fig. 6.2 Graphical representation of a decision tree. Nodes with their associated splitting test are shown as (*blue*) circles and terminal nodes with their purity output values are shown as (*green*) leaves. An event (observation) defined by variables $\vec{x}_i$ of which $H_T < 242$ GeV and $m_{\text{top}} > 162$ GeV will return $D(\vec{x}_i) = 0.82$, and an event with variables $\vec{x}_j$ of which $H_T \geq 242$ GeV and $p_T \geq 27.6$ GeV will have $D(\vec{x}_j) = 0.12$. All nodes continue to be split until they become leaves

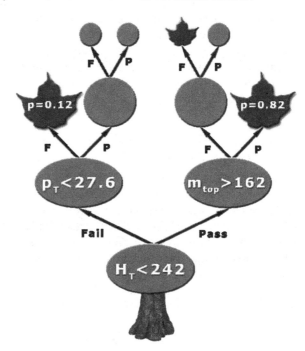

6.2.3 Advantages and Limitations

As previously mentioned, a big advantage with decision trees compared with a selection based analysis is that events which fail an individual selection criterion will continue to be considered by the algorithm.

Compared with other multivariate techniques, decision trees have several beneficial features: the tree has a human-readable structure, making it possible to know why a particular event is labelled signal or background; learning is fast compared to neural networks; decision trees can use discrete variables directly; and, no preprocessing of input variables is necessary. In addition, unlike neural networks, decision trees are relatively insensitive to including extra variables.

Adding well-modelled variables that are not powerful discriminators does not degrade the performance of the decision tree (no additional noise is added to the system).

Limitations of decision trees include the instability of the tree structure with respect to the learning sample composition, sub-optimal performance on non-linear data, and the piecewise nature of the output. Creating decision trees using random subsets of the same samples may produce very different trees, but usually with similar separation power. The decision tree output is discrete since the only possible output values are the purities of each leaf, and the number of leaves is finite.

It is possible to overcome these limitations by creating many different trees and taking the average of their output as described in Sect. 6.5. This results in a smooth combined discriminant which performs better than any individual tree. The price one has to pay is a slower more complex algorithm. One also loses the easy interpretation of why events are classified the way they are. It should be pointed out that even if the algorithm gets slower, it is in most cases still significantly faster than neural networks.

6.3 Growing a Tree

The process in which a decision tree is created is usually referred to as *decision tree learning*, but also decision tree *training*, *building* or *growing*. We start with a learning sample $\mathcal{L}$ containing N known signal and background events. Each event j is defined by an event weight w_j, a list of variables $\vec{x}_j$ and a label $y_j \in \{S, B\}$ with value S for signal and B for background. Hence we can write $\mathcal{L} = \{(w_1, \vec{x}_1, y_1), \ldots, (w_N, \vec{x}_N, y_N)\}$.

The number of weighed signal and background events in the learning sample is given by

$$s = \sum_{\mathcal{L}} w_j \times I(y_j = S), \tag{6.1}$$

and

$$b = \sum_{\mathcal{L}} w_j \times I(y_j = B), \tag{6.2}$$

where $I(\text{statement})$ is 1 if the statement is true, and 0 if not. In this analysis, each event weight, w_j, is the product of all normalization scales and efficiency corrections for the jth event, which are derived from the event properties as described in Sect. 5.4. In this situation, s and b correspond to the predicted number of signal and background events in the dataset.

The following list outlines the steps required to create a decision tree from $\mathcal{L}$. More details are found in subsequent sections as indicated in each step.

1. (Optional) Initially normalize the learning sample such that the weighted sums of signal and background become the same $(s = b)$:

$$\sum_{\mathcal{L}} w_j \times I(y_j = S) = \sum_{\mathcal{L}} w_j \times I(y_j = B). \qquad (6.3)$$

2. Create the root node with index $t = 1$ (see Fig. 6.1) containing all events in the learning sample: $\mathcal{L}_1 = \mathcal{L}$.
3. Check if any of the stopping conditions are met (see Sect. 6.4.1). If so, the node becomes a leaf and the algorithm is aborted.
4. For each variable, find the splitting value that gives the best signal-background separation (more on this in Sect. 6.3.1). If no split that improves the separation is found, the node becomes a leaf.
5. The variable and split value giving the best separation are selected, and the events $\mathcal{L}_t$ in the node are divided into two subsamples $\mathcal{L}_{2t}$ and $\mathcal{L}_{2t+1}$ depending on whether they pass or fail the split criterion. These subsamples define two new daughter nodes.
6. Apply the algorithm recursively from Step 3 until all remaining nodes have been turned into leaves.

Each leaf is assigned an output value. In most cases, the output value is the signal purity

$$p_l = \frac{s_l}{s_l + b_l}, \qquad (6.4)$$

where s_l (b_l) is the weighted sum of the signal (background) events which reach the leaf. This is the decision tree output, O_{DT}, for a given event ending in leaf l.

A leaf l is deemed a signal or background leaf depending on whether the purity p_l is greater or smaller than a parameter called the purity limit, p_{lim}. Often, this parameter is set to the initial signal purity of the sample, which is 0.5 if signal is normalized to background in Step 1 above. Each leaf l is hence associated with a class: S if $p_l \geq p_{lim}$ or B if $p_l < p_{lim}$. It is also common to define the leaf output value depending on the class only, for instance, 1 for a signal leaf, and 0 (or -1) for a background leaf.

6.3.1 Node Splitting

The most important part of decision tree learning is the *node splitting*. Here, all events $\mathcal{L}_t$ in the node t that are about to be split, are divided into the two subsets $\mathcal{L}_{2t}$ and $\mathcal{L}_{2t+1}$, which define two new daughter nodes indexed $2t$ and $2t + 1$ (see Fig. 6.1), or more simply, L (R) for the left (right) subnode. Since this is a split, the weighted sum of signal and background events in the samples are conserved: $s = s_L + s_R$ and $b = b_L + b_R$.

The best split maximizes some figure of merit (FOM) calculated from the weighted sum of signal and background events after the split. The mathematical formulation generally used is to define an impurity measure $i(s,b)$ and calculate the figure of merit for a split as the decrease in impurity, Δi:

$$\Delta i = i(s,b) - i_{\text{split}}(s_R, s_L, b_R, b_L) = i(s,b) - (i(s_L, b_L) + i(s_R, b_R)). \qquad (6.5)$$

This quantity is also referred to as the "goodness" or "gain" of the split. The split that best separates signal from background (according to the figure of merit) is the one that reduces the impurity the most (largest Δi). This split will result in the smallest impurity $i_{\text{split}}(s_R, s_L, b_R, b_L)$, since the initial condition (s, b and $i(s,b)$) is the same for all splits. Finding the best split is hence a minimization problem.

For certain applications, an alternative definition of i_{split} is used [8, 9]:

$$i_{\text{split}}(s_R, s_L, b_R, b_L) = \min(i(s_L, b_L),\, i(s_R, b_R)), \qquad (6.6)$$

which means that the right hand side of Eq. 6.5 needs to be modified accordingly.

Technically, the splitting of a node containing N events with weighted signal and background sum s and b, can be implemented in the following way:

1. Set the variable index $k = 1$, and set $s_L = b_L = 0$.
2. Sort (re-index) all events in increasing order according to the kth variable x^k. We now have $x_j^k \leq x_{j+1}^k$ for every event j.
3. Go through the events in order, and add the weight of the current event j to s_L (b_L) if the event is a signal (background) event. If it is possible to split the sample between the current event and the next ($x_j^k \neq x_{j+1}^k$), calculate Δi for the split using $s_R = s - s_L$ and $b_R = b - b_L$. Record the corresponding selection criterion: $x^k < (x_j^k + x_{j+1}^k)/2$, as the best if Δi is the highest encountered so far.
4. Set the next variable in the list as the current: $k = k + 1$, and repeat from Step 2 until all variables have been processed.
5. Split the sample according to the best split (from Step 3) if the Δi improvement for this split is positive.

6.3.2 Impurities

There are several different impurity measures suggested in literature. The two most commonly used are the Gini Index [7] and the Cross Entropy [1] defined by

$$\text{Gini Index:} \quad \frac{sb}{s+b} \qquad (6.7)$$

$$\text{Cross Entropy:} \quad -s \log \frac{s}{s+b} - b \log \frac{b}{s+b}. \qquad (6.8)$$

Both of these functions are maximal for equal amounts of signal and background and symmetric and strictly concave for any deviation thereof. It should be pointed out that most literature defines these quantities scaled by an additional factor of $(s + b)^{-1}$. Using such impurity definitions one needs to add additional factors of $s + b$ to Eq. 6.5. In this thesis, the "already weighted" impurity definitions above (also used in [5]) are used since these quantities are additive and easy to work with.

Another quantity that can be used as an impurity measure is the weighted sum of misclassified events:

$$\text{Misclassification error, } e: \quad \begin{cases} s, & \text{if } s/(s+b) < p_{\text{lim}} \\ b, & \text{otherwise.} \end{cases} \tag{6.9}$$

If $p_{\text{lim}} = 0.5$ (often the case), the definition simplifies to $e = \min(s, b)$. The misclassification error e is used for many of the traditional decision tree applications to measure the performance. For instance, if a decision tree is used to classify whether a patient is sick or not, then it is most likely desirable to have a minimal misclassification rate. As will be discussed in greater detail in Sect. 6.7, for high energy physics applications, we are usually more interested in optimizing signal significance. The following three impurity definitions have been constructed to optimize figures of merits used in high energy physics:

$$s/\sqrt{s+b}: \quad -\frac{s}{\sqrt{s+b}} \tag{6.10}$$

$$\text{Cross section significance,} \, \mathcal{S}_{\sigma}^2: \quad -\frac{s^2}{s+b} \tag{6.11}$$

$$\text{Excess significance,} \, \mathcal{S}_{s}^2: \quad -\frac{s^2}{b}. \tag{6.12}$$

The former was used in Ref. [9]; the latter two were developed and tested in this analysis.

6.4 Pruning the Tree

During the decision tree learning process described in Sect. 6.3, the crucial part is the calculation of the maximal decrease in impurity Δi. Due to the finite number of events in the learning sample, there will always be a statistical uncertainty associated with this calculation. Since the sample size is reduced after each split, the relative statistical uncertainty grows as the learning process proceeds. As a result, the splits get successively more affected by statistical fluctuations, which, in most cases, eventually leads to a degradation of performance.

To mitigate this, one usually applies so-called *pruning* criteria which limits the growth of the tree. There are two main approaches, *pre-pruning* (often just referred to as "stopping condition"), which is applied during the learning phase, and *post-pruning* (often just "pruning"), which is applied in a separate stage after the learning process is finished. The following two sections will discuss these approaches.

6.4.1 Pre-Pruning

Pre-pruning refers to one or several stopping criteria applied during the learning process (see Step 3 in Sect. 6.3). One option is to require a minimal impurity improvement for each split. The disadvantage with this method is that one might miss out on good splits that would have occurred later.

The most common approach (also used in this thesis) is to require a minimum number of events in each leaf. The idea is that if leaves are not allowed to become too small, then splits that are not statistically significant are avoided and there is little or no need for post-pruning. However when dealing with weighted events, this is not always true as discussed in Sect. 6.8

6.4.2 Post-Pruning

The idea behind post-pruning is to first grow the tree very large, and then prune the tree by turning an internal node into a leaf and hence remove the sub tree above this node. There are a long list of different pruning algorithms available. Two of the most common such methods are *Cost Complexity Pruning* and *Reduced Error Pruning*.

6.4.2.1 Reduced Error Pruning

This method was developed by Quinlan [10]. It is a recursive leaves-down method (meaning that we start from the leaves and recursively move down towards the root node). A separate pruning sample $\mathcal{P}$ is used to calculate the classification error rate of the tree. This sample needs to be independent of (orthogonal to) the learning sample: $\mathcal{P} \cap \mathcal{L} = \{\}$. For each internal node t, the number of classification errors e_t (Eq. 6.9) of the node is compared with the sum of classification errors of all the leaves in the subtree rooted at node t. The current node is pruned if the subtree has a larger error.

6.4.2.2 Cost Complexity Pruning

This method, also know as weakest link pruning or the CART pruning algorithm [1], assigns—as the name suggests—a cost for complexity. The algorithm has two stages. The first stage is a root-up recursive algorithm which creates a set of subtrees of the original tree T_{max}: $\{T_0, T_1, \ldots, T_L\}$. The crucial quantity calculated here is

$$\alpha(t) = \frac{R(t) - R_{\text{sub}}(t)}{N_{\text{leaves}}(t) - 1}, \tag{6.13}$$

where $R(t)$ (the "resubstitution estimate") is a figure of merit set by the user calculated from the events in the node (s_t and b_t), $R_{\text{sub}}(t)$ is the same quantity but calculated from all leaves in the subtree rooted at t, and N_{leaves} is the number of leaves of this subtree. In most cases, the resubstitution estimate is set to the misclassification rate $R(t) = e_t/(s_t + b_t)$. The weakest link of the tree is the node with the minimal $\alpha(t)$. The tree is pruned at this node, and the resulting tree is labelled T_j. The algorithm then repeats the same procedure starting from the root node in tree T_j. The weakest link is again found and pruned resulting in tree T_{j+1}. Eventually we end up with a tree only consisting of the root node.

In the second stage of the algorithm, all trees are evaluated using an independent pruning sample. The best performing tree is chosen, and the other trees discarded. The figure of merit used to measure the performance is often the misclassification error of the tree.

6.5 Forests of Decision Trees

This section will discuss a few algorithms that grow many decision trees and combine them into a stronger classifier. Each tree will classify an event ("vote" for its class) based on its features $\vec{x}$, and the combined output will be an average of all trees ("vote by majority").

These methods are not restricted to decision trees. Any set of weak classifiers, meaning a classifier performing slightly better than random guessing, can be combined according to these procedures. The performance (strength) of the combined discriminant depends on the strength of the individual classifiers (stronger is better), and on their correlation. The strategy behind the methods described in this section is to create a set of independent trees by forcing the learning process for each tree to emphasize a certain subset of the information available. The combined performance might improve as long as the new trees contain some degree of uncorrelated information.

Other advantages with these methods are that the learning process gets more stable and the combined discriminant output gets smoother compared to a single decision tree.

6.5.1 Bagging

Bootstrap **aggregating** is a method proposed by Breiman in 1994 [11]. The learning sample $\mathcal{L}$ is split into N_{trees} bootstrap samples $\{\mathcal{L}_1, \ldots, \mathcal{L}_{N_{\text{trees}}}\}$ by sampling randomly $f_{bs}N_L$ events from L, where $f_{bs} \leq 1$ is parameter set by the user (often $f_{bs} = 1$). The same event is allowed to be picked several times ("sampling with replacement"). Statistical fluctuations are hence introduced, randomly giving more weight to certain events. If $f_{bs} = 1$, then on average, 63.2% of the events in any of the bootstrap samples are unique, and the rest duplicated. The events which are not picked can be used to form an independent testing sample $\mathcal{T}_k$ that can be used to evaluate the performance. A decision tree (or any other classifier) is generated for each bootstrap sample $\mathcal{L}_i$, and the bagged decision tree output is the average of the output of each of the N_{trees} decision trees.

6.5.2 Random Forest

Random Forest is an extension of bagging. Just as in case of bagging, a decision tree is grown from each bootstrap sample $\mathcal{L}_k$, but an additional step is added to the learning process: when splitting a node, only the fraction f_{rf} of the N_{vars} variables is considered for the split. These $f_{rf}N_{vars}$ variables are selected randomly at each node.

6.5.3 Boosting

The idea behind boosting is to boost (assign a higher weight to) a subset of the learning sample rather than randomly selecting a subset as in case of the bagging and random forest algorithms. Many different boosting algorithms have been developed over the years. The analysis described in this thesis uses a boosting method known in the literature as AdaBoost [12]. This algorithm creates N_{trees} decision trees $\{T_1, T_2, \ldots, T_{N_{trees}}\}$ in succession, where the learning process for each tree is adapted depending on the performance of the previous tree (adaptive boosting). Once a tree T_n has been created, the events in the learning sample that are misclassified by the tree are assigned a higher weight (boosted). When creating the next tree T_{n+1}, the learning process will hence focus more on the previously misclassified events. The algorithm works as follows:

1. A first tree T_1 ($n = 1$) is created using the full learning sample $\mathcal{L}$.
2. The misclassification rate ϵ_n for the tree is calculated as the weighted sum of misclassified events in each leaf (Eq. 6.9) divided by the initial weighted sum of all events in the learning sample. A tree weight α_n is calculated according to

$$\alpha_n = \beta \times \ln\frac{1 - \epsilon_n}{\epsilon_n}, \tag{6.14}$$

where β is the boosting parameter.

3. Each misclassified event j in the learning sample is scaled by the factor e^{α_n} (which will be greater than 1): $w_j \rightarrow w_j \times e^{\alpha_n}$.
   Hence misclassified events will get higher weights.
4. (Optional) The learning sample is normalized such that the total weighted sum of signal and background events is the same as before the boosting described in the previous step. This prevents the average decision tree output value from shifting as the boosting proceeds.
5. A new tree, indexed $n + 1$, is created from the boosted sample. The learning process will now work harder on the previously misclassified events. The algorithm continues from Step 2 until N_{trees} decision trees have been created.
6. The final boosted decision tree result for event j is

$$D(\vec{x}_j) = \frac{1}{\sum_{n=1}^{N} \alpha_n} \sum_{n=1}^{N} \alpha_n D_n(\vec{x}_j), \qquad (6.15)$$

where $D_n(\vec{x}_j)$ is the decision tree output for event i from tree T_n.

An example of how the misclassification rate and the tree weights develop during the boosting procedure is illustrated in Fig. 6.3.

6.6 Decision Tree Options

Most of the different parameters and options discussed in the previous sections are summarized below.

Impurity measure figure of merit used to define the optimal split during the learning process.

Splitting condition the definition of i_{split}—most commonly according to Eq. 6.5, but Eq. 6.6 or other definitions might be useful for certain applications.

Minimal leaf size, $\mathbf{N_{leaf}^{min}}$ pre-pruning condition. The lower the value, the larger the tree.

Number of trees, $\mathbf{N_{trees}}$ the number of trees in the forest of decision trees. Applies to bagging, random forest and boosting. For boosting, $N_{trees} - 1$ boosting cycles will be applied (the first tree is created from the unboosted learning sample).

Bootstrap fraction, $\mathbf{f_{bs}}$ (Bagging and random forest only) The fraction of events to be sampled from the learning sample when creating the bootstrap samples used to create the decision trees.

Random forest variable fraction, $\mathbf{f_{rf}}$ (random forest only) Determines how many randomly selected variables that are considered when splitting each node during the random forest learning process.

AdaBoost parameter, $\boldsymbol{\beta}$ scale factor that affects the strength of the boosting. In the original algorithm, this parameter is set to unity. Lower values ($\beta \approx 0.5$ or less) often perform better. The lower the β, the softer the boosting, and the more N_{trees} might be needed to reach optimal performance.

Resubstitution estimate, $\mathbf{R(t)}$ figure of merit used during Cost Complexity Pruning.

Fig. 6.3 Example of the misclassification rate ϵ_n (*top*) and the corresponding tree weight α_n (*bottom*) versus the tree index n ($\beta = 0.2$). The misclassification rate for the individual trees tends to get worse the more boosting cycles are applied, and the tree weights hence get lower according to Eq. 6.14. Even if the individual trees perform worse, the combined performance becomes better than using the first, best tree alone as can be seen in Fig. 6.4

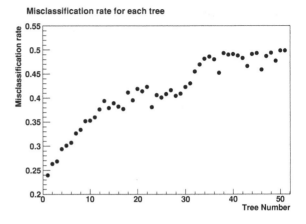

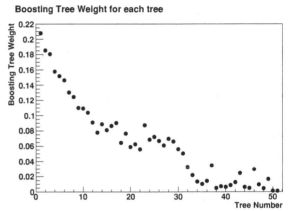

6.7 Evaluating the Performance

Since the decision tree learning process is affected by statistical fluctuations in the learning sample as discussed in Sect. 6.4, the performance evaluated on the learning sample will always be artificially enhanced. As a consequence, the performance of a decision tree must be evaluated on a (testing) sample, $\mathcal{T}$, which is completely (statistically) independent of the learning sample: $\mathcal{T} \cap \mathcal{L} = \{\}$.

A common figure of merit traditionally used to evaluate the performance of a decision tree is the misclassification rate of the testing sample. In high energy physics applications, this is not generally the quantity of interest. When trying to isolate a physics process, the goal is usually to maximize the *signal significance*. In particle physics, several definitions of significance have been used [13]. The simplest one is the ratio of expected signal excess over the statistical uncertainty of the predicted number of events, which can be written as $s/\sqrt{b}$ or $s/\sqrt{s+b}$, depending on whether we assume the background-only hypothesis, or the $s + b$ hypothesis.

In this analysis, two figures of merit where constructed: *cross section significance* and *excess significance*. As the names imply, the former is correlated to the precision of a cross section measurement, and the latter to the significance of a signal excess over background. These quantities are calculated from histograms containing the decision tree output of the testing sample separately for signal and background (see the histograms shown in Appendix C for an example). When filling these histograms, it should be ensured that the relative statistical uncertainties on the signal and background predictions are reasonably small. If this is not the case, histogram bins need to be merged, which is further discussed in Sect. 7.1.4.

6.7.1 Cross Section Significance, $\mathcal{S}_\sigma$

The cross section significance is calculated by adding the $s/\sqrt{s+b}$ significance in quadrature for each of the histogram bins of the decision tree output distribution:

$$\mathcal{S}_\sigma \equiv \sqrt{\sum_i \frac{s_i^2}{s_i + b_i}}. \tag{6.16}$$

s_i and b_i are the signal and background predictions (weighted number of events) in histogram bin i. Each histogram bin is hence treated as an individual analysis with its own dataset, similar to what is done during the actual cross section measurement described in Sect. 7.2. This analysis find that the cross section significance is a good approximation of the actual significance for the expected cross section measurement when no systematics are included in the calculation. The cross section significance $\mathcal{S}_s$ can hence be used as a quick estimate of the expected precision of a cross section measurement.

Figure 6.4 shows the cross section significance evaluated after including different numbers of boosted decision trees. It should be pointed out that only statistical uncertainties are considered when calculating $\mathcal{S}_\sigma$ according to Eq. 6.16. It should be possible to extend the formula to also include approximate systematic uncertainties, possibly as suggested in Sect. 6.8.3.

6.7.2 Excess Significance, $\mathcal{S}_s$

The excess significance is calculated by adding the $s/\sqrt{b}$ significance for each histogram bin in quadrature

$$\mathcal{S}_s \equiv \sqrt{\sum_i \frac{s_i^2}{b_i}}. \tag{6.17}$$

Since only the background prediction appears in the denominator, it is particularly important to ensure that the relative statistical uncertainty on b is reasonably small.

Figure 6.4 shows a comparison between the cross section and excess significances evaluated after combining different numbers of boosted decision trees. It is
clear that $\mathcal{S}_s$ is always larger than $\mathcal{S}_\sigma$ as expected from the definitions above. It can
also be seen how the excess significance is less stable compared to the cross
section significance since this quantity is more sensitive to statistical fluctuations
in the denominator.

6.8 Thoughts and Suggested Improvements

This section presents some of the author's thoughts of and suggested adaptations to
the decision tree algorithm to better suit the needs and applications of high energy
physics. Most of what is discussed in this section has not been tested due to time
constraints.

Fig. 6.4 Cross section
significance (*top*) and excess
significance (*bottom*) versus
number of combined decision
trees. These plots are made
from the same forest of
boosted decision trees as are
used to create the plots in
Fig. 6.3. These significance
estimates are correlated with
the expected significance
measurements described in
Sect. 7.4

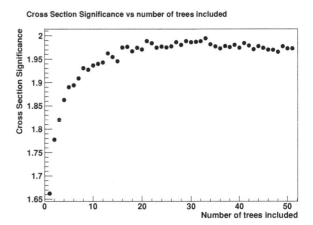

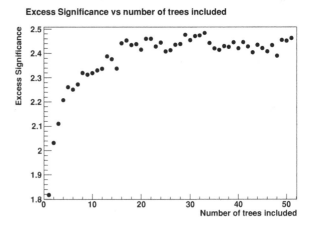

6.8.1 Weighted Events and Pre-Pruning

Most decision tree applications require a minimum number of events in each leaf. The idea is to avoid splits of leaves with very few events where statistical fluctuations will dominate the decision. This might be a reasonable approach when all learning sample events have the same weight since the relative statistical uncertainty on a count can be estimated by $1/\sqrt{N}$ (according to Poisson statistics).

In the vast majority of high energy physics applications, the input sample consists of weighted events. The sum of weighted events d in any subset of the sample corresponds to the expected number of data events in the subset: $d = \sum_{j=1}^{N} w_j$. The statistical uncertainty on the sum d can be calculated as

$$\delta d = \sqrt{\sum_{j=1}^{N} w_j^2}, \tag{6.18}$$

where N is the number of weighted events in the subset. From this uncertainty, one can define the effective number of events as the prediction squared over the statistical uncertainty of the prediction squared:

$$N_{\text{eff}} = \frac{d^2}{(\delta d)^2} = \frac{\left(\sum_{j=1}^{N} w_j\right)^2}{\sum_{j=1}^{N} w_j^2}. \tag{6.19}$$

The relative statistical uncertainty is $1/\sqrt{N_{\text{eff}}}$, and $N_{\text{eff}} = N$ when all weights are the same.

The most commonly used pre-pruning condition is a requirement on $N_{\text{leaf}}^{\text{min}}$—the minimal number of events in each leaf. When dealing with weighted input samples, it is more appropriate to instead use the number of effective events as discussed above. The requirement should be applied both to signal and background since both these quantities enter the optimization calculation (Eq. 6.5). Setting $N_{\text{leaf}}^{\text{min}} = 20$ should hence require the effective number of signal and background events in each leaf t, calculated according to Eq. 6.19 as $(s_t/\delta s_t)^2$ and $(b_t/\delta b_t)^2$, to both be greater than 20.

6.8.2 Impurity Optimization

The decision tree learning process should ideally optimize the figure of merit of interest for the analysis. A search for a new physics process could hence use the excess significance both for the learning and testing. The cross section significance should be a better choice when the goal is to measure the cross section or any other property of the signal process.

If the aim of the analysis is to apply a single selection criterion on the decision tree output distribution, as in case of particle identification, then it should be more optimal to produce the decision trees using the asymmetric node splitting criterion given by Eq. 6.6 instead of the standard splitting criterion 6.5.

6.8.3 Consideration of Systematic Uncertainties

Systematic uncertainties are an important part of all experimental particle physics analyses, and should ideally be taken into account during the decision tree learning and evaluation processes. However, it is not obvious how this information should enter the calculations without significantly increasing the learning process computing time. It is possible that the impact of the systematic uncertainties can be estimated simply by adding all uncertainties in quadrature:

$$\mathcal{S}_s^{\text{all}} \equiv \sqrt{\sum_i \frac{s_i^2}{b_i + (\delta b_i)^2 + \sum_j \epsilon_{ij}^2 b_i^2}}, \qquad (6.20)$$

where δb_i is the Monte Carlo statistics uncertainty (Eq. 6.18) and ϵ_{ij} is the relative systematic uncertainty from source j of the background in subset (bin) i. However, this does not take into account the correlations of the systematic uncertainties between the subsets (histogram bins).

6.8.4 Logging

The idea behind logging is simple: decision trees that do not improve the performance according to some figure of merit are removed. After all trees are created, the performance is first evaluated using only the first tree. Trees are then sequentially added, and the performance is reevaluated after adding each tree. If the performance degrades when adding a tree, the tree is discarded.

For instance, in Fig. 6.4, the first drop in excess significance occurs after tree number 6 is included. This would be the first tree that logging would remove (if the excess significance is the figure of merit used for logging).

References

1. L. Breiman, J. Friedman, C.J. Stone, R.A. Olshen, *Classification and Regression Trees* (Wadsworth, Stamford, 1984)
2. J.R. Quinlan, Induction of decision trees. Mach. Learn. **1**, 81–106 (1986)
3. V.M. Abazov et al.,(D0 Collaboration), it Evidence for Production of Single Top Quarks and First DirectMeasurement of |V_tb|. Phys.Rev. Lett. **98** 181802 (2007)

4. V.M. Abazov et al (dzero Collaboration),it Evidence for production of single top quarks,Phys.Rev. D 78, 012005 (2008).
5. B.P. Roe, H.-J. Yang, J. Zhu, Y. Liu, I. Stancu, G. McGregor, Boosted decision trees as an alternative to artificial neural networks for particle identification. Nucl. Instrum. Methods Phys. Res., Sect. A **543**, 577 (2005)
6. H.-J. Yang, B.P. Roe, J. Zhu, Studies of boosted decision trees for MiniBooNE particle identification, Nucl. Instrum. Methods Phys. Res., Sect. A **555**, 370 (2005)
7. C. Gini, *Variabilità e Mutabilità* (1912), reprinted in *Memorie di Metodologica Statistica*, ed. by E. Pizetti, T. Salvemini (Libreria Eredi Virgilio Veschi, Rome, 1955)
8. I. Narsky, StatPatternRecognition: A C++ package for statistical analysis of high energy physics data arXiv.org:0507.143v1 (2005)
9. I. Narsky, Optimization of signal significance by Bagging decision trees arXiv.org: 0507.157v1 (2005)
10. J.R. Quinlan, Simplifying decision trees. Int. J. Man Mach. Stud. **27**, 221–234 (1987)
11. L. Breiman, Bagging predictors, Mach. Learn. **26**, 123–140; 553–568 (1996)
12. Y. Freund, R.E. Schapire, Experiments with a new boosting algorithm, in *Machine Learning: Proceedings of the Thirteenth International Conference*, ed. by L. Saitta (Morgan Kaufmann, San Fransisco, 1996), p. 148
13. S. Bityukov, Signal significance in the presence of systematic and statistical uncertainties arXiv.org:0207.130v4 (2002)

Chapter 7
Analysis: Measurements

This chapter describes how boosted decision trees are created and applied in order to separate single top quark events from background events, and how the single top quark production cross section is measured using the boosted decision tree output distributions. Measurements of the signal significance and the CKM matrix element $|V_{tb}|$ are also presented, as well as several cross checks of the measurements.

7.1 Decision Tree Analysis

This section describes the procedure to create the boosted decision trees used in the subsequent sections. The decision tree software used was the classifier package [1] in the DØ CVS code repository. This program was originally created by Toby Burnett and Gordon Watts. Several alterations and new features were introduced mainly by Yann Coadou and the author to suit the needs of this and the previous [2, 3] analyses.

7.1.1 Input Samples

Each of the signal and background samples, created as described in Sect. 5.4, are divided into three independent subsets. The first subset of events is used for the decision tree learning, the second subset is used for decision tree optimization (Sect. 7.1.3), and the third independent subset is used for the final measurements and to produce all plots.

Because of the b-tagging modeling described in Sect. 5.4.4, all Monte Carlo samples contain permuted events with highly correlated kinematics. In order to make the subsets independent, it is important to ensure that all permutations of an event end up in the same subset. To ensure this, the samples are divided based on

D. Gillberg, *Discovery of Single Top Quark Production*, Springer Theses, 79
DOI: 10.1007/978-1-4419-7799-1_7, © Springer Science+Business Media, LLC 2011

Table 7.1 Sample splitting procedure in order to avoid bias from permuted events

Sample subset splitting procedure	
Subset	Splitting criterion
Learning subset, $\mathcal{L}$	EventNumber mod 3 = 0
Testing subset, $\mathcal{T}$	EventNumber mod 3 = 1
Yield subset, $\mathcal{Y}$	EventNumber mod 3 = 2

The event number is given to a MC event during generation and is the same for all permutations of an event. The learning subset is used to create the decision trees, the testing subset is used for decision tree evaluation and optimization, and the yield subset is used for the measurements

the modulus of the event number, which is given to Monte Carlo events during generation, as specified in Table 7.1.

This splitting procedure results in three subsets of very similar sizes. For each signal and background component, a normalization scale factor is applied to each subset such that the total sum of weights becomes the same as before splitting.

7.1.2 Discriminating Variables

One of the most important parts of a decision tree analysis is the identification and selection of the input variables. To maximize the performance, it is desirable to include uncorrelated variables with good discriminating power. However, it is crucial that all variables and their correlations are well modeled such that the decision tree performance for the modeled samples reflect the real performance in data.

Introducing more variables when creating a decision tree does not degrade the performance. If the newly introduced variables have some additional discriminative power, they will improve the performance of the tree. If they are not discriminative enough, they will be ignored. However, to reduce computing time and memory consumption, and to keep the analysis simple, it is preferred to use a reasonably short list of variables.

A long list of candidate input variables was considered for this analysis. Many of these variables were derived based on an analysis of the signal and background Feynman diagrams [4, 5] and on a study of single top quark production at next-to-leading order [2]. Other variables were constructed and evaluated for this analysis. All variables considered fall into five categories, which are described below.

- *Object kinematics* Transverse momentum p_T, pseudorapidity η, and $Q(\ell) \times \eta$ for the individual objects in the event. The latter quantity takes advantage of the CP symmetry in t-channel production as discussed in Sect. 2.3.4.
- *Event kinematics* These variables are calculated from the four-vectors of all, or a subset of the objects in the event. H, H_T and centrality are defined as

$$H = \sum_{\text{objects}} E \quad \text{energy sum,} \tag{7.1}$$

$$H_T = \sum_{\text{objects}} p_T \quad \text{scalar } p_T \text{sum}, \tag{7.2}$$

$$\text{Centrality} = H_T/H. \tag{7.3}$$

All other variables in this category are calculated from the four vector sums of the objects, for instance the invariant mass $M = \sqrt{E^2 - \vec{p}^2}$ and the transverse mass $M_T = \sqrt{E_T^2 - \vec{p}_T^2}$, where $(E, \vec{p}) = \sum_{\text{object } i}(E_i, \vec{p}_i)$. $\sqrt{\hat{s}}$ is the invariant mass of all objects in the event.

- *Angular correlations* These are either ΔR or $\Delta \phi$ angles between jets and leptons, or cosine of angles between various objects in different reference frames that have been shown to be correlated with the top quark spin [4, 5].
- *Top quark reconstruction* There are several ways to reconstruct the top quark in an event depending on which jet is used, and which neutrino solution is picked when reconstructing the W boson.

  – *Neutrino p_z solutions* The p_z of the neutrino cannot be directly measured in the detector, but can be estimated using the lepton momentum $\vec{p}(\ell)$ and the W boson mass constraint. This leads to a quadratic equation with two solutions for $p_z(v)$. In this analysis, the default choice is the solution with the smaller absolute value. However, some variables use the second solution ($S2$), with larger absolute value of $p_z(v)$.
  – *Top mass difference ΔM_{top}* The top quark mass is reconstructed for each of all possible combinations of the lepton, each neutrino solution and each jet. For any given such (ℓ, v, jet)-system, the top quark mass is calculated from

$$M_{\text{top}} = \sqrt{\left(E_\ell + E_v + E_{\text{jet}}\right)^2 - \left(\vec{p}_\ell + \vec{p}_v + \vec{p}_{\text{jet}}\right)^2}. \tag{7.4}$$

The difference between this value and 170 GeV, which is the top mass used in the Monte Carlo simulations, is called ΔM_{top}. The reconstructed top mass and ΔM_{top} for the (ℓ, v, jet)-combination yielding the smallest ΔM_{top}, define the two variables $M_{\text{top}}^{\Delta M^{\text{min}}}$ and $\Delta M_{\text{top}}^{\text{min}}$ respectively.
  – *Significance of top quark candidate* In addition to calculating the mass difference, the significance of the reconstructed top mass, Significance (M_{top}), is also calculated for each (ℓ, v, jet) combination. This quantity relates the mass difference ΔM_{top} with the resolution of the reconstructed top mass. It is assumed that the top quark mass resolution is a Gaussian distribution of width δM_{top}, and the significance of the reconstructed top mass is calculated from

$$\text{Significance } (M_{\text{top}}) = \ln \frac{\text{Gauss } (\Delta M_{\text{top}}/(\delta M_{\text{top}}))}{\text{Gauss } (0)}, \tag{7.5}$$

where Gauss is the probability density function of a Gaussian with mean 0 and width 1. The resolution uncertainty of the reconstructed top mass δM_{top} is derived in terms of the resolution of the $\not{E}_T, \delta \not{E}_T$ [6], and the jet energy resolution, δE_{jet} [7], by error propagation of 7.4. The lepton energy resolution is neglected as the lepton energy is well measured compared to that of the jets and $\not{E}_T$. The variables Significance$_{min}$ (M_{top}) and M_{top}^{sig} are defined as the significance and reconstructed top mass from the $(\ell, \nu, \text{ jet})$ combination that gives the smallest top mass significance in an event.

• *Jet reconstruction* The jet width in η and ϕ is the energy weighted root-mean-square of the η and ϕ for all cells in the jet energy cluster.

Starting from several hundred variables, the variable list is reduced in two steps: variables that showed unsatisfactory data-background agreement are removed; and the most sensitive variables are identified and selected.

To judge whether a variable is well modeled or not, the variable distribution for data is compared with the sum of the signal and backgrounds. This is done for each of the 24 channels individually. Two requirements are enforced. The Kolmogorov–Smirnov test value [3], calculated by comparing the variable distribution for data with the sum of signal and backgrounds, is required to be at least 0.1 for the majority of the channels. Then the data-background agreement has to be judged as satisfactory after examining the data-background agreement by eye.

In order to further reduce the list, the most discriminating variables are identified using decision tree variable ranking. This ranking is obtained by creating one or several decision trees and for each variable calculating the sum of impurity improvements Δi for each split in which the variable is used. Hence frequently used variables tend to get high decision tree rankings, while an unused variable will get a ranking equal to zero. Decision trees are created using the full list of well modeled variables for each channel individually. A combined list of variables is created using the 50 highest ranked variables from the 2jets,1tag channels, the 30 best from the 3jets, 1tag channels, the 20 best from the 2jets, 2tag channels, and the 10 best from each of the other three Njets, Ntags combinations. After removing the duplicated entries in this combined list, a final list of 64 well-modeled variables is obtained.

The 64 variables are listed in Table 7.2, and the data-background agreement for all channels combined can be seen in Figs. 7.1, 7.2, 7.3, 7.4, 7.5 and 7.6. The variable name describes which objects are included when calculating the variable value. Jets are sorted in p_T, and index 1 refers to the leading jet in a jet category. "jet$n$" ($n = 1, 2, 3, 4$) corresponds to each jet in the event. "tag$n$" are the b-tagged jets, "light$n$" are defined as all jets but the leading b-tagged jet. The "best" jet is the one for which the invariant mass $M(W, \text{jet})$ is closest to $m_{top} = 170$ GeV, and "notbest$n$" are all but the best jet.

7.1.3 Choice of Decision Tree Parameters

There are many parameters that can impact the performance of a decision tree (see Sect. 6.6). The impact from various parameter choices is studied by creating several sets of decision trees, using the learning subset of events as described in

Table 7.2 The 64 variables used as input to the decision trees, in five categories: object kinematics, jet reconstruction, angular correlations, event kinematics, and top quark reconstruction

Decision tree input variables	
Object kinematics	Event kinematics
$p_T(\text{jet2})$	Centrality(alljets)
$p_T(\text{jet3})$	$H_T(\text{alljets})$
$p_T(\text{jet4})$	$H_T(\text{alljets}-\text{tag1})$
$p_T(\text{tag1})$	$H_T(\text{alljets}-\text{best})$
$p_T(\text{light2})$	$H_T(\text{jet1, jet2})$
$p_T(\text{notbest2})$	$H_T(\text{jet1, jet2, lepton}, \not{E}_T)$
$p_T(\text{lepton})$	$H_T(\text{alljets, lepton}, \not{E}_T)$
$\not{E}_T$	$H_T(\not{E}_T, \text{lepton})$
$Q(\text{lepton}) \times \eta(\text{jet1})$	$H(\text{alljets}-\text{tag1})$
$Q(\text{lepton}) \times \eta(\text{jet2})$	$M(\text{alljets})$
$Q(\text{lepton}) \times \eta(\text{best})$	$M(\text{alljets}-\text{best})$
$Q(\text{lepton}) \times \eta(\text{light1})$	$M(\text{alljets}-\text{tag1})$
$Q(\text{lepton}) \times \eta(\text{light2})$	$M(\text{jet1, jet2})$
	$M(\text{jet1, jet2, } W)$
Jet widths	$M(\text{jet3, jet4})$
$\text{Width}_\eta(\text{jet2})$	$M_T(\text{jet1, jet2})$
$\text{Width}_\eta(\text{jet4})$	$p_T(\text{jet1, jet2})$
$\text{Width}_\phi(\text{jet4})$	$\sqrt{\hat{s}}$
$\text{Width}_\eta(\text{tag1})$	$M_T(W)$
$\text{Width}_\eta(\text{light2})$	
$\text{Width}_\phi(\text{light2})$	
Angular correlations	Top quark reconstruction
$\Delta R(\text{jet1, jet2})$	$M(W, \text{best1})$ ("best" top mass)
$\Delta R(\text{jet1, lepton})$	$M(W, \text{tag1})$ ("b-tagged" top mass)
$\Delta R(\text{tag1, lepton})$	$M(W, \text{tag1}, S2)$ (with second neutrino solution)
$\Delta R(\text{light1, lepton})$	$M(W, \text{jet1})$
$\Delta\phi(\text{lepton}, \not{E}_T)$	$M(W, \text{jet1}, S2)$
$\cos(\text{best, lepton})_{\text{besttop}}$	$M(W, \text{jet2})$
$\cos(\text{best, notbest})_{\text{besttop}}$	$M(W, \text{jet2}, S2)$
$\cos(\text{jet1, lepton})_{\text{btaggedtop}}$	$M(W, \text{notbest2})$
$\cos(\text{tag1, lepton})_{\text{btaggedtop}}$	$M(W, \text{notbest2}, S2)$
$\cos(\text{lepton}_{\text{besttop}}, \text{besttop}_{\text{CMframe}})$	$M_{\text{top}}^{\Delta M^{\min}}$
$\cos(\text{lepton}_{\text{btaggedtop}}, \text{btaggedtop}_{\text{CMframe}})$	$M_{\text{top}}^{\text{sig}}$
$\cos(\text{tag1, lepton})_{\text{btaggedtop}}$	$\Delta M_{\text{top}}^{\min}$
$\cos(\text{lepton}, Q(\text{lepton}) \times z)_{\text{besttop}}$	$\text{Significance}_{\min}(M_{\text{top}})$

For the angular variables, the subscript indicates the reference frame

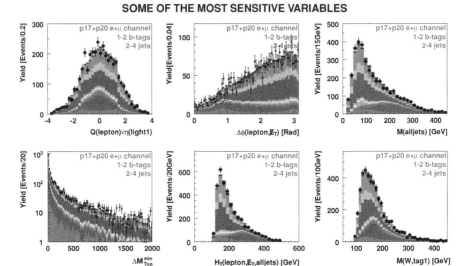

Fig. 7.1 Distributions of some of the most sensitive variables used as input when creating the boosted decision trees. The plot key can be seen in Fig. 5.6

Sect. 7.1.1, and then evaluating the performance using the testing subset. The figure of merit used to evaluate the performance is the *cross section significance*, (6.16) calculated using histograms with 100 bins with equal widths.

The parameter choice from the previous boosted decision tree analysis [2, 5] is used as a starting point. Several different parameters are varied one at a time over the range of values shown in Table 7.3. The strategy is to identify the optimal parameter value, fix this value and optimize the next parameter. In order to reduce the computational time, the evaluation is only done using two of the 24 channels, namely the Run IIb, 2jets, 1tag e and μ channels, which are two of the most sensitive channels.

7.1.3.1 Boosting Parameters

The results from varying the AdaBoost parameter and the number of boosting cycles are shown in Fig. 7.7. The result clearly improves when going from 0 to 20 boosts, thereafter the performance improves only marginally and reaches a plateau. This can also be seen with better resolution in Fig. 6.4.

The largest AdaBoost parameter value $\beta = 0.5$ performs worse than the other choices. β values around 0.2 perform equally well within uncertainty.

7.1.3.2 Impurity Measures

The boosted decision tree performance for three different impurity measures are shown versus different number of boosting cycles in Fig. 7.8. The performance of

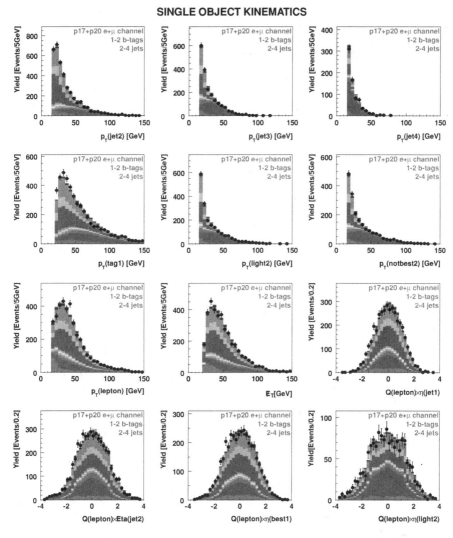

Fig. 7.2 Distributions for most individual object kinematic variables used as input to the decision trees. The plot key can be seen in Fig. 5.6

$\mathcal{S}_s$ (excess significance, (6.12) is not shown since it performed significantly worse than the others. The reason for this is most likely statistical instability of the background estimation b in the denominator. This problem might be solved by requiring a minimal number of effective background events in each leaf as discussed in Sect. 6.8.1. The performances for the other impurity measures does not differ significantly.

EVENT KINEMATICS

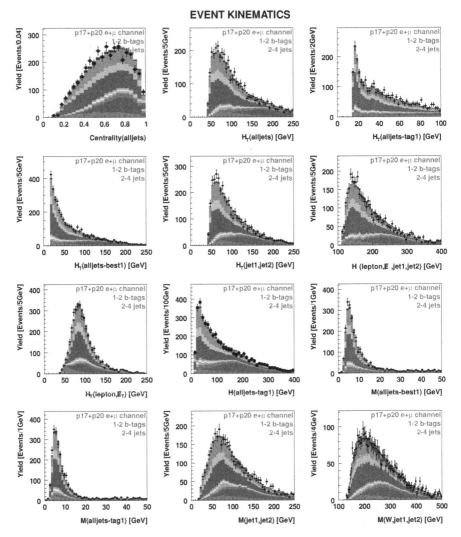

Fig. 7.3 Distributions for event kinematic variables used as input to the decision trees. The plot key can be seen in Fig. 5.6

7.1.3.3 Minimal Leaf Size

Figure 7.9 shows the decision tree performance for different minimal leaf size values. A smaller minimal leaf size results in a larger tree. Leaf sizes in the range 50–200 all perform equally within uncertainty.

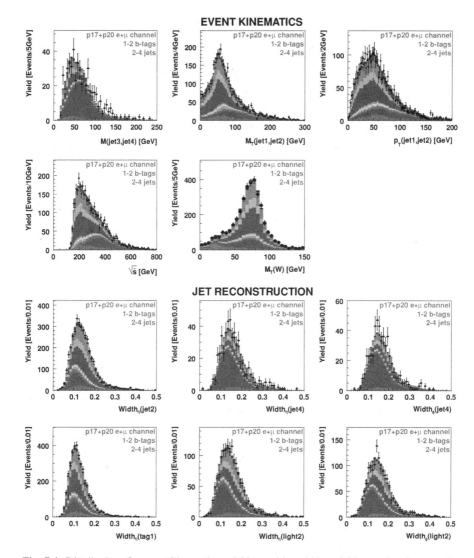

Fig. 7.4 Distributions for event kinematic variables and jet width variables used as input to the decision trees. The plot key can be seen in Fig. 5.6

7.1.3.4 Pruning

Two different pruning methods are tested: Cost complexity pruning and reduced error pruning. No improvement in performance is observed using any of these methods (compared to no pruning). Hard pruning resulted in worse performance, softer pruning made no difference. This is probably because the pre-pruning choice of $N_{\text{leaves}}^{\text{min}} = 100$ results in a close to optimally grown tree which needs no further pruning.

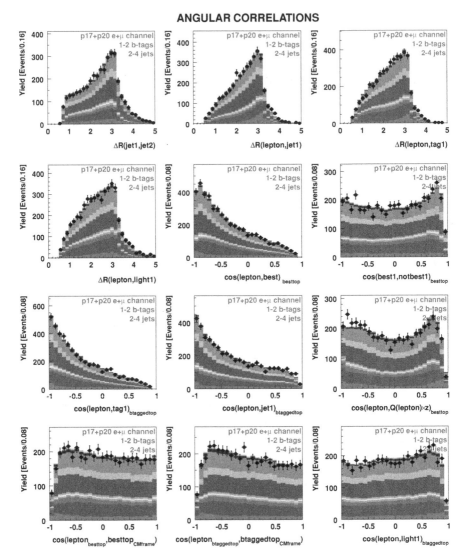

Fig. 7.5 Distributions for most angular correlation variables used as input to the decision trees. The plot key can be seen in Fig. 5.6

7.1.3.5 Summary

From the study described above, the parameters listed in Table 7.4 are chosen. This list of parameter settings results in a good separation for the channels studied: (Run IIa, e+2jets, 1tag) and (Run IIa, μ+2jets, 1tag). The same set of parameters is used for the other analysis channels since no significance differences of optimal parameters are expected, and since most of these channels have significantly less impact on the final result.

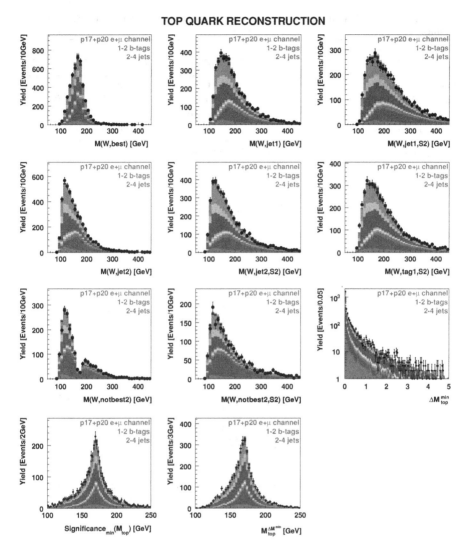

Fig. 7.6 Distributions for most top quark reconstruction variables used as input to the decision trees. The plot key can be seen in Fig. 5.6

7.1.4 Output Transformation

The decision tree output distribution given by (6.15) tends to be very sparsely populated close to 0 and close to 1 as can be seen in the top plot of Fig. 7.10. This results in problems with the stability of the cross section calculation since the signal and background estimations in some histogram bins are based on very few simulated events, and hence have a large Monte Carlo statistical uncertainty. This

Table 7.3 The decision tree parameter values that are evaluated during the decision tree optimization

Boosted decision tree parameter scan	
Parameter	Evaluation points
Impurity measure	**Gini**, Entropy, $\mathcal{S}_\sigma$, $\mathcal{S}_s$
Minimal leaf size, $N_{\text{leaf}}^{\text{min}}$	50, 75, 90, **100**, 110, 125, 150, 200, 500
Number of boosting cycles, N_{boosts}	0, **20**, 30, 50, 70
AdaBoost parameter, β	0.05, 0.15, 0.18, **0.20**, 0.22, 0.25, 0.3, 0.5

The parameter values used in the previous single top analysis are indicated with bold font

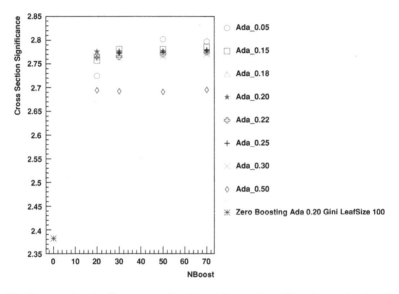

Fig. 7.7 Cross section significance as a function of the number of boosting cycles for different values of the AdaBoost parameter. The statistical uncertainty on the evaluation points is about ±0.05

is particularly troublesome in the signal region, where many bins have very few background events. In extreme cases, there might even be bins containing signal events but no background events.

A monotonic re-binning scheme was designed in order to remove the instability described above without losing too much resolution in the signal region. The re-binning is done individually for each channel, and transforms the background distribution such that it follows a $1/x$-curve up to 0.8, and a linear slope from the intercept of the $1/x$ graph at 0.8, down to zero at 0.95. There are no shape constraints between 0.95 and 1.0, but all bins of width 0.02 are required to contain at least 40 background events in order to keep the statistical uncertainty reasonably small. The transformation is done from right $(O_{BDT} = 1.0)$ to left.

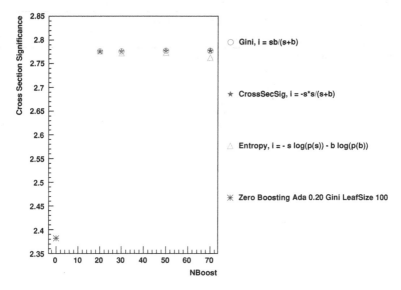

Fig. 7.8 Cross section significance as a function of the number of boosting cycles for different impurity measures used by the decision tree learning process. The uncertainty on the evaluation points are about ±0.05

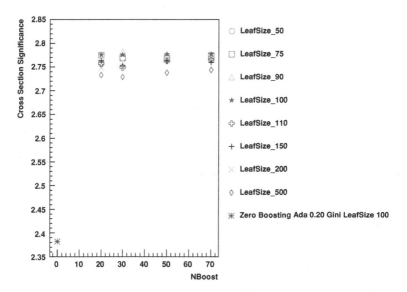

Fig. 7.9 Cross section significance as a function of the number of boosting cycles for different minimum leaf size values. The uncertainty on each cross section significance point is roughly ±0.05

Table 7.4 The decision tree parameter values that are chosen based on the procedure described in Sect. 7.1.3

Chosen decision tree parameters	
Parameter	Value
Impurity measure	Gini
Minimal leaf size, $N_{\text{leaf}}^{\text{min}}$	100
Number of boosting cycles, N_{boosts}	50
AdaBoost parameter, β	0.20

These parameters are used when creating all final decision trees

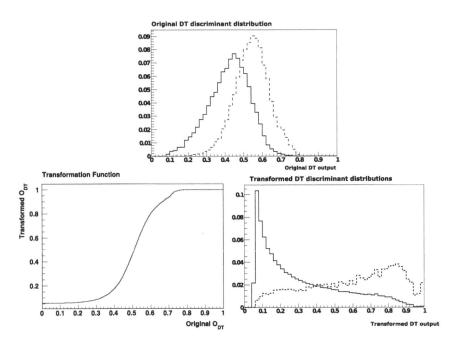

Fig. 7.10 The original boosted decision tree discriminant for signal dashed line and all background components combined solid line (*top*). The distributions are sparsely populated in the high and low discriminant regions. In this particular histogram there is one histogram bin in the signal region containing only three background events. The monotonic transformation function (*bottom left*) is applied to both the signal and the background. The resulting transformed boosted decision tree distributions (*bottom right*) have adequate statistics in all histogram bins. These plots are for the Run IIa, e+jets, 2 jets, 1 tag channel

The transformed background shape is

$$f_1(x) = k/x, \qquad \text{when } x < 0.8,$$
$$f_2(x) = M - Kx, \quad \text{when } 0.8 < x < 0.95,$$

with $k = 0.346, K = 2.88$ and $M = 2.74$, such that $f_1(0.8) = f_2(0.8)$.

Technically, a histogram of the original boosted decision tree distribution is created with 10,000 uniform bins between 0 and 1 and filled with the weights of all background events. The histogram is normalized to unity. A new, initially empty histogram with 50 uniform bins is defined. Starting from the right ($O_{BDT} = 1$) in the original histogram, the content of each bin is moved to the rightmost bin in the new histogram until the two conditions mentioned above are met: enough background statistics and enough background events to make the weighted sum equal or greater than the value expected from the function. This procedure continues until the contents of all histogram bins have been moved to the new histogram. Each bin in the old histogram now has a corresponding bin in the new histogram, and a "transfer function" can be derived from this mapping (see Fig. 7.10).

This rebinning is equivalent to creating a histogram with variable bin-widths of the native boosted decision tree distribution. The width of each bin is then given by the conditions described above.

7.1.5 The Final Decision Trees

The final boosted decision trees are created using the learning subset for each of the 24 channels separately. The variable list presented in Table 7.2 and the decision tree parameters in Table 7.4 are used for all channels. Various properties of the 24 boosted decision trees are presented in Table 7.5. The average size of the trees is related to the number of events in the input sample, which increases with the number of jets and b-tags due to the permuted events.

The boosted decision tree output distributions for each of the 24 individual boosted decision trees are shown in Appendix C in Figs. C.1–C.4. Figure 7.11 presents all these distributions combined, by stacking the histograms. This does not truly reflect the performance since each channel is considered individually when measuring the cross section. Figure 7.12 shows the boosted decision tree distributions for the six different (Njet, Ntag) combinations after combining the Run IIa and IIb and the e and μ channels.

7.1.6 Cross Checks

In order to validate every step of the decision tree analysis without being biased by a potential sign of signal, cross-check samples are created and used to decide whether the background model and data are in agreement after applying the boosted decision trees. The selection criteria "W+jets": (2 jets, 1 tag, $H_T < 175$ GeV) and "$t\bar{t}$": (4 jets, 1 or 2 tags, $H_T > 300$ GeV) are applied to construct samples dominated by the W+jets and $t\bar{t}$ backgrounds. Figure 7.13 shows the decision tree output distributions in these cross-check samples for Run IIa-b, e and μ, 1–2 tags combined. In Appendix D, the distributions for the individual channels are shown separately for Run IIa and Run IIb, for e+jets and μ+jets.

Table 7.5 Various properties for the 24 boosted decision trees

Properties for the 24 boosted decision trees

Channel	$\langle N_{\text{nodes}} \rangle$	$\langle N_{\text{leaves}} \rangle$	$\langle$depth$\rangle$	S_σ	S_s
Run IIa, e+2jets, 1tag	596	299	20	2.00	2.48
Run IIa, e+3jets, 1tag	280	141	9	0.95	1.06
Run IIa, e+4jets, 1tag	190	96	9	0.45	0.49
Run IIa, e+2jets, 2tags	343	172	19	1.05	1.21
Run IIa, e+3jets, 2tags	394	197	15	0.69	0.81
Run IIa, e+4jets, 2tags	692	347	24	0.39	0.44
RunIIa, μ+2jets, 1tag	693	347	23	2.19	2.82
Run IIa, μ+3jets, 1tag	649	325	26	1.13	1.36
Run IIa, μ+4jets, 1tag	353	177	18	0.48	0.52
Run IIa, μ+2jets, 2tags	300	151	13	1.16	1.35
Run IIa, μ+3jets, 2tags	984	493	30	0.81	0.96
Run IIa, μ+4jets, 2tags	992	497	27	0.40	0.46
Run IIb, e+2jets, 1tag	335	168	16	1.82	2.13
Run IIb, e+3jets, 1tag	334	168	20	1.08	1.26
Run IIb, e+4jets, 1tag	178	89	11	0.55	0.61
Run IIb, e+2jets, 2tags	255	128	17	1.00	1.15
Run IIb, e+3jets, 2tags	516	258	24	0.73	0.90
Run IIb, e+4jets, 2tags	279	140	12	0.41	0.47
Run IIb, μ+2jets, 1tag	458	229	19	2.00	2.46
Run IIb, μ+3jets, 1tag	301	151	14	1.17	1.34
Run IIb, μ+4jets, 1tag	278	139	16	0.61	0.71
Run IIb, μ+2jets, 2tags	112	56	7	1.02	1.16
Run IIb, μ+3jets, 2tags	661	331	26	0.80	1.00
Run IIb, μ+4jets, 2tags	465	233	15	0.46	0.54

Each boosted decision tree contains a forest of 51 decision trees, and the average number of nodes and leaves per decision tree and the average tree depth is shown in the three first columns. These quantities are mainly related to the size of the learning sample. The cross section significance and the excess significance calculated after applying all trees are given in the last two columns

As an additional cross check, the decision trees are applied to the data and simulated samples before any b tagging is applied. The purpose of this exercise is to look at a sample composed of a large number of events and verify that the data and background are in agreement. A complication with this exercise is that the boosted decision trees are trained with several variables that use information about the b-tagged, and untagged jets (see Sect. 7.1.2). Since this information is not available at the pre b-tagging stage, variables associated with tagged, or anti-tagged jets are replaced according to Table 7.6. This change might affect the validity of the cross check, but in principle the data-background agreement should still be adequate since the decision trees treat data and background equally.

For each pre-tag subsample, the decision tree created in the corresponding 1tag-channel is applied. The resulting boosted decision tree output distributions for the four pre-tag channels with 2 jets are shown in Fig. 7.14. The corresponding distributions for all twelve pre-tag channels are shown in Fig. D.3 of Appendix D.

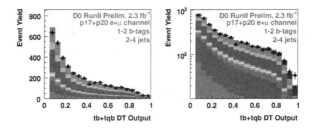

Fig. 7.11 Decision tree discriminant output for all 24 channels combined using linear scale (*left*) and log scale (*right*) for the *y*-axis. The plot colour key for the signal and background components can be seen in Fig. 5.6

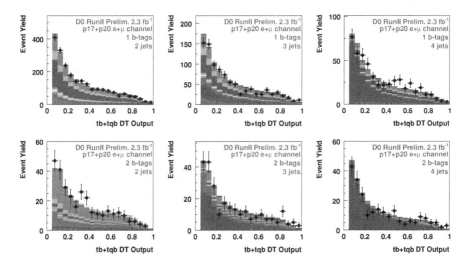

Fig. 7.12 Boosted decision tree discriminant output distributions after combining the Run IIa and Run IIb run periods as well as the *e* and *μ* channels. Combined channels with one *b*-tagged jet are shown in the *top row*, with two *b*-tagged jets in the *bottom row*, and with two, three and four jets in the *left*, *middle* and *right* columns, respectively. The plot key can be seen in Fig. 5.6

All boosted decision tree distributions shown in this section, and in Appendix D, show good agreement between data and the background model. No bias from the decision trees due to the composition of the background model is observed. Since the background model and boosted decision trees behave well, the analysis can move forward with confidence to measure the single top cross section.

7.2 Cross Section Measurement

This section describes how the single top cross section is extracted from the final observed boosted decision tree distributions seen in Figs. C.1–C.4, and how the

Fig. 7.13 Combined
decision tree outputs for the
"W+jets" sample (*left*) and
the "$t\bar{t}$" sample (*right*) cross-
check samples

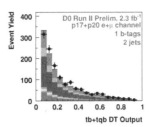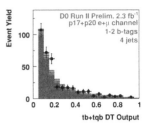

Table 7.6 The variables derived using the objects in the left column, are replaced by the corresponding variables using the information in the right column when the boosted decision trees
are applied to the pre-b-tagging (pre-tag) sample

Variable replacement for the pre-tag sample

Used for decision tree learning	Used when applying to pre-tag data
Leading b-tagged jet, *tag*1	Best-top-mass jet, *best*
Leading light-quark jet, *light*1	Leading not-best jet, *notbest1*
Second light-quark jet, *light*2	Second not-best jet, *notbest2*

This is done since no b-tagging information is available for the events in the pre-tag sample.
Further explanation of the variables and the naming convention is given in Sect. 7.1.2

measurement is cross checked and calibrated using ensembles of pseudo-data. All
measurements assume the standard model ratio of the s and t-channel single top
cross section: $\sigma_{tb}/\sigma_{tqb} = 1.12/2.34 = 0.48$ (see Table 2.3).

7.2.1 Bayesian Analysis

In a given histogram bin, the probability to observe D data events, if the expected
number of events is d, is given by the Poisson distribution

$$P(D|d) = \frac{e^{-d}d^D}{\Gamma(D+1)}, \tag{7.6}$$

where Γ is the gamma function. The expected number of events d in the bin is the
sum of the predicted signal s and background b, which further can be expressed as

$$d = s + b = a\sigma + \sum_{j=1}^{N_{\text{bkg}}} b_j, \tag{7.7}$$

where a is the effective luminosity for the signal, σ is the signal cross section, b_j is
the expected number of events (yield) of background source j and N_{bkg} is the
number of background sources. When dealing with many bins from a single or
several histograms, one can construct a combined likelihood as a product of the
single-bin likelihoods [3, 8]

Fig. 7.14 Boosted decision tree output distribution from applying the final 2jets,1tag boosted decision trees to the corresponding 2jets channels before *b* tagging is applied

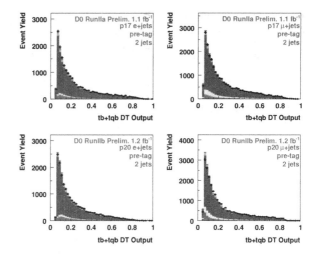

$$L(\mathbf{D}|\mathbf{d}) \equiv L(\mathbf{D}|\sigma, \mathbf{a}, \mathbf{b}) = \prod_{i=1}^{N_{\text{bins}}} P(D_i|d_i),\qquad(7.8)$$

where $\mathbf{D}$ and $\mathbf{d}$ are vectors of the observed and predicted number of events in each bin, and $\mathbf{a}$ and $\mathbf{b}$ are vectors of effective luminosity and background yields. Using Bayes' theorem, the posterior probability density $p(\sigma, \mathbf{a}, \mathbf{b}|\mathbf{D})$ can be obtained and further converted into the function of interest by integrating with respect to the parameters $\mathbf{a}$ and $\mathbf{b}$ [3, 9]:

$$p(\sigma|\mathbf{D}) = \frac{1}{\mathcal{N}} \iint L(\mathbf{D}|\sigma, \mathbf{a}, \mathbf{b}) \pi(\mathbf{a}, \mathbf{b}) \pi(\sigma) \mathrm{d}\mathbf{a} \mathrm{d}\mathbf{b}.\qquad(7.9)$$

$\mathcal{N}$ is here an overall normalization factor, the prior density $\pi(\sigma)$ is set to $1/\sigma_{\text{max}}$ for $0 < \sigma < \sigma_{\text{max}}$, and 0 otherwise. The prior probability density $\pi(\mathbf{a}, \mathbf{b})$ encodes all knowledge of the effective signal luminosity and background yields, including all systematic uncertainties and their correlations.

The peak position of the $p(\sigma|\mathbf{D})$ distribution is interpreted as the measured cross section, and the 68% interval around the peak as the uncertainty of the measurement as illustrated in Fig. 7.15. This interval is constructed such that the posterior probabilities are equal at the start and end points of the interval.

7.2.2 Numerical Calculation

The integration of (7.9) is done numerically using Monte Carlo sampling. N_{samples} systematically-shifted histograms $(\mathbf{a}_k, \mathbf{b}_k)$ are generated by random sampling from the prior density $\pi(\mathbf{a}, \mathbf{b})$. Uncertainties that affect the normalization only are modeled as the widths of Gaussian distributions with means set to the expected

Fig. 7.15 Illustration of the
posterior density $p(\sigma|\mathbf{D})$. The
measured cross section
is the peak position σ_{peak},
and the uncertainty of the
measurement is the interval
$\Delta\sigma$ covering 68.27% of the
posterior as indicated in the
plot

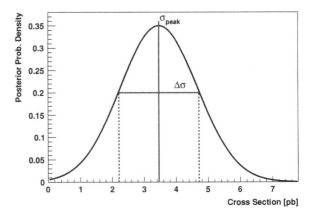

yields. Systematic uncertainties that affect the boosted decision tree discriminant
shape are modeled bin by bin by evaluating the effect of shifting the uncertainty up
and down by one standard deviation. This results in different positive and negative
shifts as illustrated in Fig. 7.16. This is further discussed in Appendix B. Uncertainties that are correlated between different bins, are treated by using the same
random Gaussian shift.

Using the systematically-shifted histograms $(\mathbf{a}_k, \mathbf{b}_k)$, the posterior density given
in (7.9) is estimated by

$$p(\sigma|\mathbf{D}) \approx \frac{1}{\mathcal{N}\sigma_{\text{max}}N_{\text{samples}}} \sum_{k=1}^{N_{\text{samples}}} L(\mathbf{D}|\sigma, \mathbf{a}_k, \mathbf{b}_k). \tag{7.10}$$

For the combined measurements in this analysis, the posterior density is calculated
using this formula with σ_{max} set to 12 pb, and N_{samples} set to 20,000.

7.2.3 Ensemble Tests

Ensemble testing is performed in order to ensure that there is no bias in the cross
section measurements. An ensemble is a collection of pseudo-data sets generated
with a known signal to background fraction. Each pseudo-data set is randomly
sampled from the signal and background in the yield sample (see Sect. 7.1.1),
taking into account both statistical and systematical uncertainties. The probability
to pick any given event is proportional to the event weight modified by the unique
systematic shifts for the pseudo-dataset in question. The pseudo data hence mimic
all expected characteristics of real data, and can also be treated just like real data.

Eight ensembles are generated with the single top cross section set to 2, 3, 3.46
(standard model), 4.2, 5, 7, 8 and 10 pb, respectively. When generating these
ensembles, the event weights for the single top events are initially scaled such that

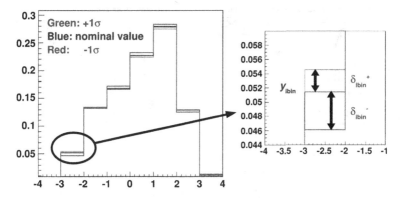

Fig. 7.16 Illustration of the treatment of shape shifting systematic uncertainties. This analysis uses shape-changing systematics for the jet energy scale, b-tagging efficiency and the ALPGEN reweighting. Separate boosted decision tree histograms are created from events where these quantities are shifted up and down by one standard deviation (see Sect. B.2). The systematic uncertainty in any given bin is modeled by a Gaussian distribution with different positive and negative widths, δ_{ibin}^{+} and δ_{ibin}^{-}, as illustrated in the plot

the probability to sample a single top event is increased or decreased by the desired amount.

The cross section is measured for each pseudo-data set, treating the pseudo-data in exactly the same way as real data. The measured cross sections in the ensembles can be seen in Fig. 7.17, where also a Gaussian fit is performed around the peak of the distribution. The average measured cross sections closely match the cross section used when generating the ensembles. This is further illustrated in Fig. 7.18, which shows a linear calibration fit from the means of the fitted Gaussians shown in Fig. 7.17. No correction to the cross section measurements is hence needed.

The distribution of measured cross sections in the ensemble containing the standard model amount of single top resemble the *standard model expectation* of the cross section measurement. The average measured cross section is very close to the standard model value of 3.46 pb, and the distribution has a standard deviation of 0.90 pb.

7.2.4 Observed Results

This section presents the boosted decision tree cross section measurements using the 2.3 fb^{-1} dataset. The histograms used for the cross section calculation are shown in Appendix C. The boosted decision tree output for all channels stacked on top of each other is shown in Fig. 7.19, visualizing the excess of data over background in the signal region.

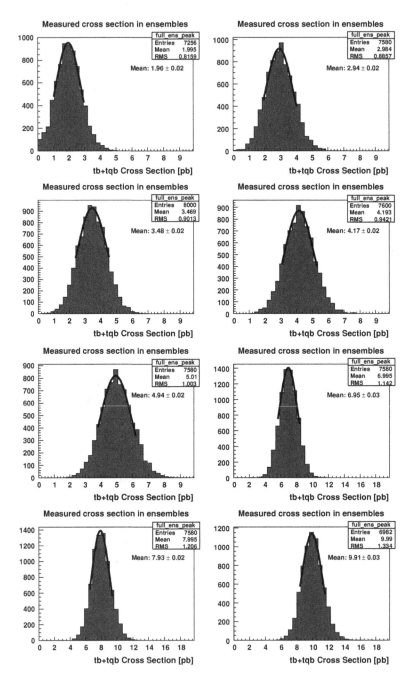

Fig. 7.17 Measured single top cross sections in ensembles generated with various amounts of single top. The input single top cross sections used are 2, 3, 3.46 (SM), 4.2, 5, 7, 8 and 10 pb

Fig. 7.18 Linear fit through the means from the Gaussian fits (Fig. 7.17) of the measured cross sections in ensembles generated with different amounts of single top. The fit is constrained to the range [2, 10]. The correct cross section is measured on average

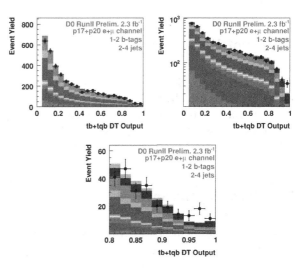

Fig. 7.19 Boosted decision tree discriminant output distributions for all 24 channels combined. The single top contribution in this plot is normalized to the measured cross section. The same combined distribution is shown on linear scale (*top left*), log scale (*top right*) and a zoom in the signal region (*below*)

The measurement is performed as described in Sect. 7.2.1, and the resulting posterior density for all channels combined is shown in Fig. 7.20. The measured cross sections is

$$\sigma(p\bar{p} \rightarrow tb + X, tqb + X) = 3.74^{+0.95}_{-0.74}\text{pb}.$$

This measurement assumes the standard model ratio of single top s and t-channel production $\sigma_{tb}/\sigma_{tqb} = 0.48$. The measured cross sections for various combinations of analysis channels are presented in Table 7.7. All results are consistent with the standard model cross section of 3.46 pb within uncertainty. The peak over half-width significance (P/HW) is defined as the ratio of posterior peak position over the lower 68.3% confidence bound. The peak over half-width values for various combinations of analysis channels are presented in Table 7.8. Table 7.9 further presents the measurements in each of the 24 individual channels.

Fig. 7.20 Observed posterior density from $s + t$-channel single top cross section measurement using boosted decision trees. This is for all 24 channels combined—i.e., Run IIa and Run IIb, e+jets and μ+jets, 2–4 jets and 1 or 2 of them b-tagged. The *lines* show the 1σ (68.3%), 3σ (99.7%) and 5σ (99.99994%) confidence bounds. All systematic uncertainties are taken into account in this measurement

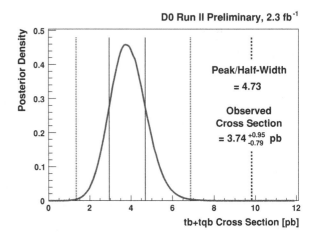

7.3 Event Kinematics

Figures 7.21 and 7.22 show data-background comparisons for various discriminating variables used by the boosted decision trees after applying different cuts on the decision tree discriminant. Single top in these plots are normalized to the measured cross section. Event displays of two of the most signal like events ($O_{\mathrm{BDT}} > 0.98$) are shown in Appendix A.

7.4 Signal Significance

The significance of the excess over background is measured using a very large ensemble of pseudo-datasets generated with background only. Each such dataset corresponds to $2.3\,\mathrm{fb}^{-1}$ of data without any single top. The single top cross section is measured in each such pseudo-dataset in exactly the same way as for the real dataset.

From the measured cross sections in the ensemble, the probability for background only to fluctuate to give a cross section higher than the standard model cross section, or the measurement in real data, is calculated. This probability is referred to as the "p-value", and is widely used to estimate the significance of a measurement. From a p-value α, the number of standard deviations equivalence N_σ is calculated using

$$N_\sigma = \sqrt{2} \cdot \mathrm{erf}^{-1}(1 - 2\alpha) \qquad (7.11)$$

which fulfils

$$\int_{-\infty}^{N_\sigma} \mathrm{Gauss}(x)dx = 1 - \alpha, \qquad (7.12)$$

where the normal distribution $\mathrm{Gauss}(x)$ is normalized to unity.

Table 7.7 Measured single top quark production cross sections for many different combinations of analysis channels

| | Observed cross section measurements | | | | | | | All channels |
| | 1, 2tags + 2, 3, 4jets | | e, μ + 2, 3, 4jets | | e, μ + 1, 2tags | | | |
	e-chan	μ-chan	1 tag	2 tags	2 jets	3 jets	4 jets	
Run IIa	$2.3^{+1.7}_{-1.6}$	$2.7^{+1.6}_{-1.5}$	$1.9^{+1.3}_{-1.2}$	$3.7^{+2.6}_{-2.3}$	$1.2^{+1.1}_{-1.0}$	$4.7^{+3.0}_{-2.7}$	$5.8^{+6.7}_{-4.6}$	$2.50^{+1.29}_{-1.16}$
Run IIb	$6.2^{+2.2}_{-1.9}$	$3.9^{+1.7}_{-1.5}$	$5.8^{+1.6}_{-1.6}$	$3.8^{+2.5}_{-2.2}$	$4.3^{+1.8}_{-1.5}$	$5.6^{+2.9}_{-2.5}$	$9.2^{+6.8}_{-5.2}$	$4.92^{+1.35}_{-1.21}$
Run IIa+b	$4.4^{+1.5}_{-1.3}$	$3.3^{+1.2}_{-1.0}$	$3.8^{+1.1}_{-0.9}$	$3.7^{+1.9}_{-1.7}$	$2.6^{+1.1}_{-1.0}$	$5.2^{+2.1}_{-1.8}$	$7.0^{+5.3}_{-3.9}$	$3.74^{+0.95}_{-0.79}$

All systematic uncertainties are taken into account in these measurements

Table 7.8 Posterior peak over half-width significance for many different combinations of analysis channels

	Observed posterior peak over half-width							All channels
	1, 2tags + 2, 3, 4jets		e, μ + 2, 3, 4jets		e, μ + 1, 2tags			
	e-chan	μ-chan	1 tag	2 tags	2 jets	3 jets	4 jets	
Run IIa	1.4	1.9	1.6	1.6	1.2	1.8	1.3	**2.2**
Run IIb	3.2	2.6	3.7	1.7	2.8	2.3	1.8	**4.1**
Run IIa+b	3.6	3.2	4.1	2.2	2.6	2.9	1.8	**4.7**

The best values from all channels combined are shown in bold type. All systematic uncertainties are taken into account in these calculations

Table 7.9 Measured cross sections and peak over half-width significances, with all systematic uncertainties taken into account, for each of the 24 individual analysis channels

Observed results in individual channels

Channel	$\sigma \pm \Delta\sigma$	P/HW
e / p17 / 1tag / 2jets	$0.91^{+1.80}_{-0.91}$	1.0
e / p17 / 1tag / 3jets	$9.03^{+6.89}_{-5.41}$	1.7
e / p17 / 1tag / 4jets	$8.15^{+11.22}_{-8.15}$	1.0
e / p17 / 2tags / 2jets	$0.00^{+3.16}_{-0.00}$	0.0
e / p17 / 2tags / 3jets	$9.27^{+7.94}_{-6.31}$	1.5
e / p17 / 2tags / 4jets	$0.00^{+15.51}_{-0.00}$	0.0
e / p20 / 1tag / 2jets	$0.00^{+0.00}_{-0.00}$	0.0
e / p20 / 1tag / 3jets	$5.65^{+4.33}_{-3.57}$	1.6
e / p20 / 1tag / 4jets	$14.15^{+12.53}_{-9.67}$	1.5
e / p20 / 2tags / 2jets	$4.24^{+4.37}_{-3.62}$	1.2
e / p20 / 2tags / 3jets	$5.04^{+6.37}_{-4.87}$	1.0
e / p20 / 2tags / 4jets	$21.37^{+19.44}_{-13.66}$	1.6
μ / p17 / 1tag / 2jets	$2.53^{+1.96}_{-1.73}$	1.5
μ / p17 / 1tag / 3jets	$0.81^{+3.42}_{-0.81}$	1.0
μ / p17 / 1tag / 4jets	$0.00^{+7.05}_{-0.00}$	0.0
μ / p17 / 2tags / 2jets	$1.56^{+3.99}_{-1.56}$	1.0
μ / p17 / 2tags / 3jets	$1.00^{+6.35}_{-1.00}$	1.0
μ / p17 / 2tags / 4jets	$12.65^{+13.95}_{-9.16}$	1.4
μ / p20 / 1tag / 2jets	$5.05^{+2.58}_{-2.19}$	2.3
μ / p20 / 1tag / 3jets	$5.19^{+4.50}_{-3.69}$	1.4
μ / p20 / 1tag / 4jets	$3.62^{+10.38}_{-3.62}$	1.0
μ / p20 / 2tags / 2jets	$2.02^{+4.19}_{-2.02}$	1.0
μ / p20 / 2tags / 3jets	$4.38^{+5.17}_{-4.02}$	1.1
μ / p20 / 2tags / 4jets	$8.68^{+11.18}_{-8.05}$	1.1

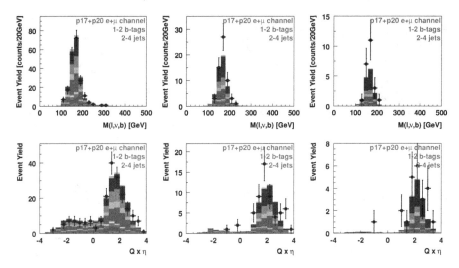

Fig. 7.21 Data-background comparison for various variables after requiring $O_{BDT} > 0.8$ (*left column*), $O_{BDT} > 0.9$ (*middle column*) and $O_{BDT} > 0.96$ (*right column*). All channels combined

Figure 7.23 presents the expected and observed significances for the signal excess over background in the boosted decision tree distribution. The expected significance, sometimes referred to as the *sensitivity*, is calculated from the fraction of pseudo-datasets measuring a cross section above the standard model single top cross section of 3.46 pb. 267 pseudo-datasets out of 34.1 million measure a single top cross above 3.46 pb, which corresponds to an expected 4.3σ excess over background. The observed significance is calculated from the number of pseudo-datasets that measure a cross section higher than the cross section measured in real data, and is hence strongly correlated with the measured cross section. The observed significance for this analysis is 4.6σ.

7.4.1 Combined Significance

Two other multivariate analyses were performed using the same data and simulated samples: one based on Bayesian neural networks (BNN) [10, 11], and one using the matrix element method (ME) [12, 13]. Just as for the boosted decision tree analysis described in this thesis, these are improved versions of the single top evidence analyses, which are described in Ref. [3].

The single top cross section and the significance are measured individually for each analysis. These measurements are highly correlated since the same dataset is used. However, since the analyses use quite different techniques to isolate single top quarks, they are not fully correlated. The three analyses were therefore combined into a more powerful discriminant using a second Bayesian neural network [14]. The discriminant outputs for the individual multivariate techniques as well as the BNN combination are presented in Appendix E.

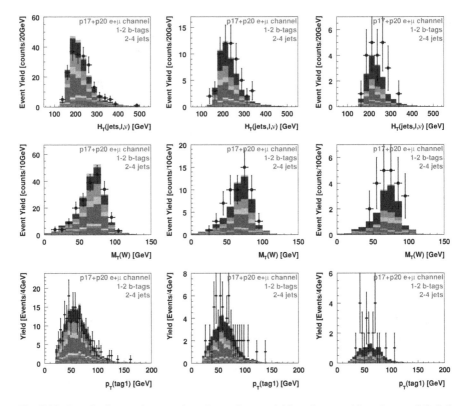

Fig. 7.22 Data-background comparison for various variables after requiring $O_{BDT} > 0.8$ (*left column*), $O_{BDT} > 0.9$ (*middle column*) and $O_{BDT} > 0.96$ (*right column*). All channels combined

The sensitivity of the combined measurement is determined using a very large ensemble of background-only pseudo-datasets, in the same way as for the boosted decision tree analysis (Sect. 7.2.3). The distribution of measured cross sections in the background-only ensemble and the expected and observed significances are presented in Fig. 7.24. The boosted decision tree and BNN combination results are summarized in Table 7.10.

The observed significance for the BNN combination exceeds 5σ, which corresponds to the first observation of single top quark production.

7.5 Measurement of $|V_{tb}|$

As discussed in Sect. 2.3.3, a measurement of the amplitude of the CKM matrix element V_{tb} can be performed from the boosted decision tree discriminant output in much the same way as the cross section measurement since the single top cross section is directly proportional to $|V_{tb}|^2$. This measurement makes no assumptions on the number of quark families or the unitarity of the CKM matrix. However,

Fig. 7.23 Measured cross sections using the boosted decision tree distributions in a large ensemble of pseudo-datasets containing no single top. The expected significance (*top*) is calculated from the number of pseudo-datasets measuring a cross section higher than the standard model cross section, and the observed significance (*bottom*) is derived from the number of pseudo-datasets with measured cross section above the measurement in real data

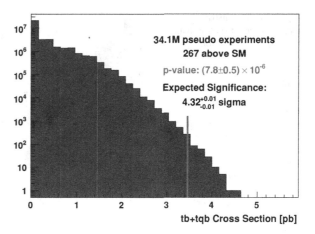

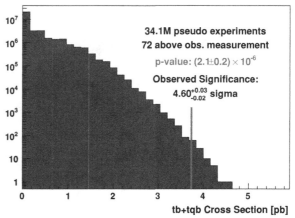

assumptions are made in the interpretation of the measurement as well as when generating the Monte Carlo samples, as discussed below.

$|V_{tb}|$ is assumed to be much larger than $|V_{td}| + |V_{ts}|$ such that $\mathcal{B}(t \to Wb) \simeq$ 100%. This assumption is used in the Monte Carlo generation and is reasonable since measurements of the quantity $R = |V_{tb}|/(|V_{tb}| + |V_{td}| + |V_{ts}|)$ are consistent with unity [15]. Single top quarks are assumed only to be produced via the standard model production modes (Sect. 2.3.2), hence no single top production via flavour-changing neutral currents or new, heavy charged bosons (Sect. 2.3.6) are considered. Finally, the *Wtb* interaction is assumed to be CP-conserving and of the *V–A* type, but is allowed to have an anomalous strength f_L. Adding this factor results in a *Wtb* vertex of the form [2, 16]

$$\Gamma_{Wtb}^{\mu} = -\frac{-ig_w}{\sqrt{2}} f_1^L V_{tb} \bar{u}(p_b) \gamma^{\mu} P_L u(p_t) \tag{7.13}$$

where $P_L = (1 - \gamma_5)/2$ is the left handed projection operator.

Fig. 7.24 The distribution of measured cross section using the BNN combination discriminant in a very large ensemble of pseudo-datasets containing no single top. The expected (*above*) and observed (*below*) significances are calculated from the number of pseudo-datasets measuring a cross section higher than the expected standard model measurement and the measurement using real data, respectively

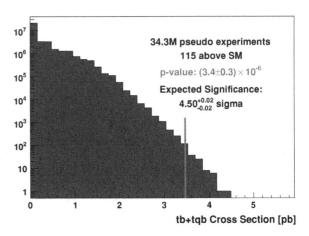

34.3M pseudo experiments
115 above SM

p-value: $(3.4\pm0.3)\times10^{-6}$

Expected Significance:
$4.50^{+0.02}_{-0.02}$ sigma

tb+tqb Cross Section [pb]

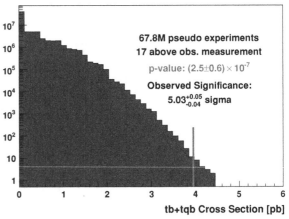

67.8M pseudo experiments
17 above obs. measurement

p-value: $(2.5\pm0.6)\times10^{-7}$

Observed Significance:
$5.03^{+0.05}_{-0.04}$ sigma

tb+tqb Cross Section [pb]

Table 7.10 Expected and observed significances and measured single top cross sections for the boosted decision tree analysis as well as for the BNN combination

Multivariate analysis results			
Analysis	Significance		Measured
	Expected	Observed	σ_{s+t}[pb]
Boosted decision trees	4.3σ	4.6σ	$3.7^{+1.0}_{-0.8}$
BNN combination	4.5σ	5.0σ	$3.9^{+0.9}_{-0.9}$

The BNN combination results in an improved expected significance, and measures an observed significance of 5.0σ

Under these assumptions, the $|V_{tb}|$ measurement is conducted using the same Bayesian calculations (Sect. 7.2.1) and the same boosted decision tree histograms as for the cross section measurement. A few additional systematic uncertainties need to be considered during this measurement. The magnitude and the sources for these systematics are shown in Table 7.11. Two measurements are performed. The first measurement is "unconstrained" (uses a flat prior between 0 and 3), and

Table 7.11 Systematic uncertainties in percent that need to be considered when measuring $|V_{tb}|$ in addition to all uncertainties affecting the cross section measurement (Sect. 5.8)

Additional systematic uncertainties in percent affecting the $|V_{tb}|$ measurement

	tb	tqb
Top quark mass	5.56	3.48
Factorization scale	3.7	1.74
PDF	3.0	3.0
α_s	1.4	0.01

Fig. 7.25 $|V_{tb}f_1^L|^2$ measurement result using an unconstrained prior. All systematic uncertainties are taken into account in this measurement, including the additional systematics listed in Table 7.11. The different shaded regions represent the 68.3, 95.4 and 99.7% confidence bounds

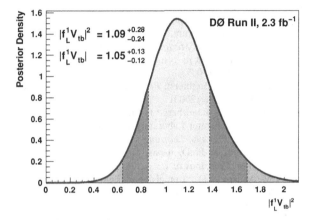

Fig. 7.26 $|V_{tb}|^2$ measurement using a flat prior between 0 and 1. All systematic uncertainties are taken into account in this measurement, including the additional systematics listed in Table 7.11. The different shaded regions represent the 68.3, 95.4 and 99.7% confidence bounds

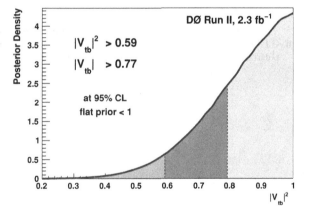

the second measurement is restricted to the $[0, 1]$ interval. The former can be interpreted as a measurement of $|f_1^L V_{tb}|$, while the latter only considers the region allowed by the standard model ($f_1^L = 1$), and can hence be interpreted as a measurement of $|V_{tb}|$.

The first measurement yields $|V_{tb}f_1^L| = 1.05^{+0.13}_{-0.12}$, and the posterior density is presented in Fig. 7.25. The second posterior density peaks at unity and is shown in Fig. 7.26. The corresponding measurement yields $|V_{tb}| = 1.00^{+0.00}_{-0.11}$, or $|V_{tb}| > 0.77$ at 95% confidence level.

References

1. The classifier decision tree package: http://cdcvs0.fnal.gov/cgi-bin/public-cvs/cvsweb-public. cgi/classifier/?cvsroot=d0cvs
2. Q.H. Cao, R. Schwienhorst, C.P. Yuan, Next-to-leading order corrections to single top quark production and decay at the Tevatron: 1: s-Channel process. Phys. Rev. D **71**, 054023 (2005)
3. D.C. Boes, F.A. Graybill, A.M. Mood, *Introduction to the Theory of Statistics*. 3rd ed. (McGraw-Hill. New York, 1974)
4. L. Dudko, Use of neural networks in a search for single top quark production at DØ *AIP Conference Proceedings* **583**, 83 (2001)
5. E. Boos, L. Dudko, Optimized neural networks to search for Higgs Boson production at the Tevatron. Nucl. Instrum. Meth. **A 502**, 486 (2003)
6. Ariel Schwartzman, *Missing Et Significance Algorithm in RunII data*, DØ Note 4254 (2003)
7. M. Voutilainen, Jet p_T resolution for Run IIa final JES (v7.2) with dijet J4S jet corrections. DØ Note 5499 (2007)
8. E.T. Jaynes, L. Bretthorst, *Probability Theory: The Logic of Science*, Cambridge University Press, Cambridge (2003)
9. I. Bertram, G. Landsberg, J. Linneman, R. Partridge, M. Paterno, H.B. Prosper, Fermi National Accelerator Laboratory Technical Memorandum No. 2104 (2000)
10. R.M. Neal, *Bayesian Learning for Neural Networks*. (Springer-Verlag, New York, 1996)
11. A. Tanasijczuk, Ph.D. thesis, *Single Top Quark Cross Section Measurement in Proton-Antiproton Collisions at $\sqrt{s} = 1.96\,TeV$.* Universidad de Buenos Aires (in preparation)
12. V.M. Abazov et al., A precision measurement of the mass of the top quark. Nature **429**, 638 (D0 Collaboration) (2004)
13. M. Pangilinan, Ph.D. thesis, Observation of Single Top Quark Production at DØ using Matrix Elements. Brown University (in preparation)
14. V.M. Abazov et al., Observation of single top quark production, arXiv:0903.0850 (D0 Collaboration) (2009)
15. V.M. Abazov et al., (D0 Collaboration), Phys. Lett. B **639**, 616 (2006)
16. G.L. Kane, G.A. Ladinsky, C.-P. Yuan, Using the top quark for testing standard-model polarization and *CP predictions*, Phys. Rev. D **45**, 124 (1992)

Appendix A

A.1 Event Displays

This appendix shows event displays of two signal candidate events. The first one is shown in Figs. A.1, A.2, and A.3 and is an e+jets event with three jets, one of them b-tagged. The second event is shown in Figs. A.4, A.5, and A.6 and is a μ+jets event, with three jets of which two are b-tagged. The boosted decision tree outputs for the two events are 0.984 and 0.991, respectively.

Run 229388 Evt 13339887 Wed Jan 3 21:05:14 2007

ET scale: 39 GeV

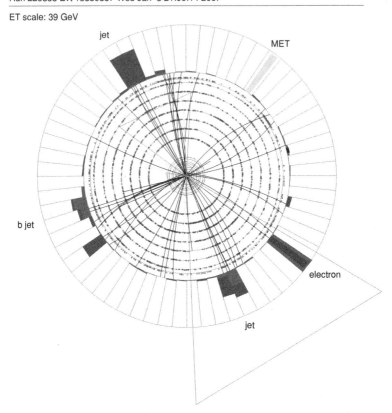

Fig. A.1 Transverse-plane view of a single *b*-tagged *e*+3 jets signal candidate event. The positive *x*-axis points to the right, and the *y*-axis points up. Hits in the inner tracking system are shown as *red dots* and *blue circles*, reconstructed tracks are shown as *black lines*, and electromagnetic and hadronic energy deposits in the calorimeter towers are illustrated as *red* and *blue bars*. The yellow bar (*top-right*) is the reconstructed missing transverse energy vector, $\vec{E}_T$, and the *dark red bar* with a matched track to the bottom-right is the electron. Other views of the same event can be seen in Figs. A.2 and A.3

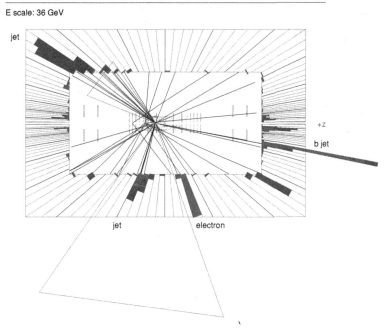

Run 229388 Evt 13339887 Wed Jan 3 21:05:14 2007

E scale: 36 GeV

Fig. A.2 Side view of the single-tagged $e+3$ jets signal candidate event show in Figs. A.4 and A.6. The z-axis points to the right, and the outer, *thin lines* are lines of constant η drawn in increments of 0.1. The *bars* illustrate energy deposits in the electromagnetic (*red*) and hadronic (*blue*) layers of the calorimeter towers. The *upper half* of the plot illustrates the positive y hemisphere $(0 < \phi < \pi)$ and the *lower half* represents the negative y hemisphere $(\pi < \phi < 2\pi)$.

Run 229388 Evt 13339887 Wed Jan 3 21:05:14 2007

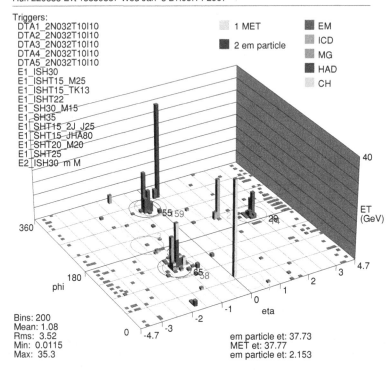

Fig. A.3 (η, ϕ) "lego plot" of the single-tagged $e+3$ jets signal candidate event shown in Figs. A.1 and A.2. The *brown bar* at $(\eta, \phi) = (0.43, 5.65)$ is the reconstructed electron with $p_T = 37.7\,\text{GeV}$, the *yellow bar* show the magnitude and the ϕ-coordinate of the $\not{E}_T$ (it is placed at $\eta = 0$ since the z-coordinate is unknown). Again, the *red* and *blue bars* show the electromagnetic and hadronic energy deposits in the calorimeter, respectively

Run 223473 Evt 27278544 Sun Jul 23 19:21:41 2006

ET scale: 28 GeV

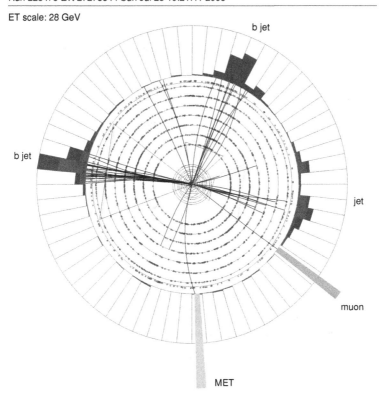

Fig. A.4 Transverse-plane view of a double b-tagged μ+3 jets signal candidate event. The x-axis is horizontal pointing right and the y-axis is vertical pointing up. Hits in the inner tracking system are shown as *red dots* and *blue circles*, reconstructed tracks are shown as *black lines* and electromagnetic and hadronic energy deposits in the calorimeter towers are illustrated as *red* and *blue bars*. The *yellow bar* (*bottom*) is the reconstructed missing transverse energy vector, $\vec{\not{E}}_T$, the *green bar* (*bottom-right*) is a muon. The jet to the right is a forward jet ($\eta = 2.2$) that is not b-tagged, the other two jets are both b-tagged. Other views of the same event can be seen in Figs. A.5 and A.6

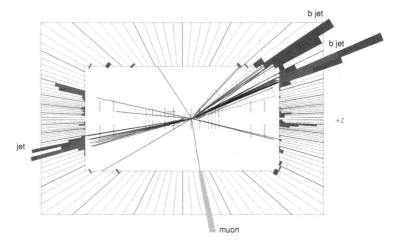

Run 223473 Evt 27278544 Sun Jul 23 19:21:41 2006

E scale: 28 GeV

Fig. A.5 Side view of the double *b*-tagged μ+3 jets signal candidate event show in Fig. A.4. The *z*-axis points to the right, and the outer, thin lines are lines of constant η drawn in increments of 0.1. The *bars* illustrate energy deposits in the electromagnetic (*red*) and hadronic (*blue*) layers of the calorimeter towers. The *upper half* of the plot illustrates the positive *y* hemisphere $(0 < \phi < \pi)$ and the lower half represents the opposite hemisphere $(\pi < \phi < 2\pi)$

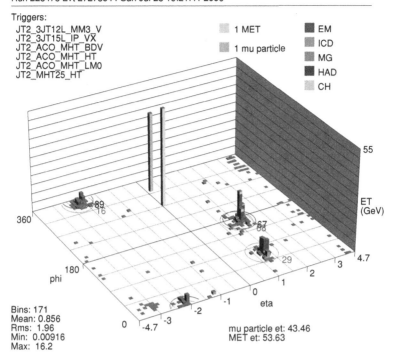

Run 223473 Evt 27278544 Sun Jul 23 19:21:41 2006

Triggers:
JT2_3JT12L_MM3_V
JT2_3JT15L_IP_VX
JT2_ACO_MHT_BDV
JT2_ACO_MHT_HT
JT2_ACO_MHT_LM0
JT2_MHT25_HT

1 MET   EM
1 mu particle ICD
               MG
               HAD
               CH

ET (GeV)
55
4.7

eta
phi
360
180
0

Bins: 171
Mean: 0.856
Rms: 1.96
Min: 0.00916
Max: 16.2

mu particle et: 43.46
MET et: 53.63

Fig. A.6 (η, ϕ) "lego plot" of the double b-tagged μ+3 jets signal candidate event show in Figs. A.4 and A.5. The *green bar* illustrates the (η, ϕ)-coordinates and the momentum of the muon, the *yellow bar* shows the magnitude and the ϕ-coordinate of the $\not{E}_T$ (it is placed at $\eta = 0$ since the z-coordinate is unknown). The *red* and *blue bars* again show the electromagnetic and hadronic energy deposits in the calorimeter, respectively

Appendix B

B.1 Systematic Uncertainties

B.1.1 Systematics Affecting Normalization Only

Tables B.1, B.2, B.3, B.4, B.5 and B.6 show the systematic uncertainties on the signal and background samples that affect the normalization only. There are also three systematics sources that affect the shapes of the distributions: jet energy scale, tag-rate functions, and ALPGEN W+jets reweighting factors. These effects are not included in the tables since they are treated differently in the calculations, and are discussed separately in Sect. B.2

The tables show the correlations between various background components and analysis channels for each uncertainty. A systematic uncertainty is assumed to be fully correlated between all signal or background samples within a given row in each table, and for rows with the same name in different tables. This does not fully apply to the lepton identification and trigger uncertainties, which are treated independently for electrons and muons and the Run IIa and Run IIb run periods.

Since the W+jets and multijets backgrounds are normalized to data before b tagging (Sect. 5.4.6), the simulated W+jets components are not affected by most of the systematic uncertainties. However, there are uncertainties on the relative compositions of the W+jets components, and due to the W+jets and multijets normalization. These uncertainties are anticorrelated due to the constraint to match data before b-tagging, which is indicated by giving one of the values a negative sign. It should also be pointed out that there is a normalization uncertainty due to the b tagging of the simulated samples, which is not shown in the tables. This uncertainty is roughly 7 and 11% for events with one and two b-tagged jets respectively (see Figs. B.1, B.2, B.3, B.4).

Table B.1 Systematic uncertainties for the Run IIa electron channels with two jets

| | Single-tagged two-jet electron channel percentage uncertainties | | | | | | | | | | |
	$t\bar{t}$	$Wb\bar{b}$	$Wc\bar{c}$	Wlp	$Zb\bar{b}$	$Zc\bar{c}$	Zlp	Dibosons	Multijet	tb	tqb
Luminosity	6.1	–	–	–	6.1	6.1	6.1	6.1	–	6.1	6.1
Xsect.	12.7	–	–	–	5.8	5.8	5.8	5.8	–	11.2	7.4
Branching frac.	1.5	–	–	–	–	–	–	–	–	1.5	1.5
PDF	–	–	–	–	–	–	–	–	–	3.0	3.0
Triggers	5.0	–	–	–	5.0	5.0	5.0	5.0	–	5.0	5.0
Lumi. rewtg.	1.0	–	–	–	1.0	1.0	1.0	1.0	–	1.0	1.0
Prim. vertex	1.4	–	–	–	1.4	1.4	1.4	1.4	–	1.4	1.4
Lepton ID	2.5	–	–	–	2.5	2.5	2.5	2.5	–	2.5	2.5
Jet frag.	0.7	–	–	–	4.0	4.0	4.0	0.7	–	0.7	0.7
ISR/FSR	3.0	–	–	–	8.0	8.0	8.0	0.6	–	0.6	0.6
b-jet frag.	2.0	–	–	–	2.0	–	–	–	–	2.0	2.0
Jet ID	1.0	–	–	–	1.0	1.0	1.0	1.0	–	1.0	1.0
Jet res.	4.0	–	–	–	4.0	4.0	4.0	4.0	–	4.0	4.0
S_{HF}^W	–	13.7	−1.5	13.7	–	–	–	–	–	–	–
S_{HF}^{ratio}	–	−5.0	–	5.0	–	–	–	–	–	–	–
S_{HF}^Z	–	–	–	–	13.7	13.7	–	–	–	–	–
IKS	–	2.3	2.3	2.3	–	–	–	–	−42.0	–	–

| | Double-tagged two-jet electron channel percentage uncertainties | | | | | | | | | | |
	$t\bar{t}$	$Wb\bar{b}$	$Wc\bar{c}$	Wlp	$Zb\bar{b}$	$Zc\bar{c}$	Zlp	Dibosons	Multijet	tb	tqb
Luminosity	6.1	–	–	–	6.1	6.1	6.1	6.1	–	6.1	6.1
Xsect.	12.7	–	–	–	5.8	5.8	5.8	5.8	–	11.2	7.4
Branching frac.	1.5	–	–	–	–	–	–	–	–	1.5	1.5
PDF	–	–	–	–	–	–	–	–	–	3.0	3.0
Triggers	5.0	–	–	–	5.0	5.0	5.0	5.0	–	5.0	5.0
Lumi. rewtg.	1.0	–	–	–	1.0	1.0	1.0	1.0	–	1.0	1.0
Prim. vertex	1.4	–	–	–	1.4	1.4	1.4	1.4	–	1.4	1.4
Lepton ID	2.5	–	–	–	2.5	2.5	2.5	2.5	–	2.5	2.5
Jet frag.	0.7	–	–	–	4.0	4.0	4.0	0.7	–	0.7	0.7
ISR/FSR	3.0	–	–	–	8.0	8.0	8.0	0.6	–	0.6	0.6
b-jet frag.	2.0	–	–	–	2.0	–	–	–	–	2.0	2.0
Jet ID	1.0	–	–	–	1.0	1.0	1.0	1.0	–	1.0	1.0
Jet res.	4.0	–	–	–	4.0	4.0	4.0	4.0	–	4.0	4.0
S_{HF}^W	–	13.7	−1.5	13.7	–	–	–	–	–	–	–
S_{HF}^{ratio}	–	−5.0	–	5.0	–	–	–	–	–	–	–
S_{HF}^Z	–	–	–	–	13.7	13.7	–	–	–	–	–
IKS	–	2.3	2.3	2.3	–	–	–	–	−42.0	–	–

The shape-shifting systematic uncertainties are not shown in these tables (see Sect. B.2)

Table B.2 Systematic uncertainties for the Run IIa electron channels with three jets

	$t\bar{t}$	$Wb\bar{b}$	$Wc\bar{c}$	Wlp	$Zb\bar{b}$	$Zc\bar{c}$	Zlp	Dibosons	Multijet	tb	tqb
	Single-tagged three-jet electron channel percentage uncertainties										
Luminosity	6.1	–	–	–	6.1	6.1	6.1	6.1	–	6.1	6.1
Xsect.	12.7	–	–	–	5.8	5.8	5.8	5.8	–	11.2	7.4
Branching frac.	1.5	–	–	–	–	–	–	–	–	1.5	1.5
PDF	–	–	–	–	–	–	–	–	–	3.0	3.0
Triggers	5.0	–	–	–	5.0	5.0	5.0	5.0	–	5.0	5.0
Lumi. rewtg.	1.0	–	–	–	1.0	1.0	1.0	1.0	–	1.0	1.0
Prim. vertex	1.4	–	–	–	1.4	1.4	1.4	1.4	–	1.4	1.4
Lepton ID	2.5	–	–	–	2.5	2.5	2.5	2.5	–	2.5	2.5
Jet frag.	0.1	–	–	–	4.0	4.0	4.0	3.7	–	3.7	3.7
ISR/FSR	2.8	–	–	–	8.0	8.0	8.0	5.2	–	5.2	5.2
b-jet frag.	2.0	–	–	–	2.0	–	–	–	–	2.0	2.0
Jet ID	1.0	–	–	–	1.0	1.0	1.0	1.0	–	1.0	1.0
Jet res.	4.0	–	–	–	4.0	4.0	4.0	4.0	–	4.0	4.0
S_{HF}^{W}	–	13.7	−0.8	13.7	–	–	–	–	–	–	–
S_{HF}^{ratio}	–	−5.0	–	5.0	–	–	–	–	–	–	–
S_{HF}^{Z}	–	–	–	–	13.7	13.7	–	–	–	–	–
IKS	–	1.8	1.8	1.8	–	–	–	–	−30.0	–	–
	Double-tagged three-jet electron channel percentage uncertainties										
	$t\bar{t}$	$Wb\bar{b}$	$Wc\bar{c}$	Wlp	$Zb\bar{b}$	$Zc\bar{c}$	Zlp	Dibosons	Multijet	tb	tqb
Luminosity	6.1	–	–	–	6.1	6.1	6.1	6.1	–	6.1	6.1
Xsect.	12.7	–	–	–	5.8	5.8	5.8	5.8	–	11.2	7.4
Branching frac.	1.5	–	–	–	–	–	–	–	–	1.5	1.5
PDF	–	–	–	–	–	–	–	–	–	3.0	3.0
Triggers	5.0	–	–	–	5.0	5.0	5.0	5.0	–	5.0	5.0
Lumi. rewtg.	1.0	–	–	–	1.0	1.0	1.0	1.0	–	1.0	1.0
Prim. vertex	1.4	–	–	–	1.4	1.4	1.4	1.4	–	1.4	1.4
Lepton ID	2.5	–	–	–	2.5	2.5	2.5	2.5	–	2.5	2.5
Jet frag.	0.1	–	–	–	4.0	4.0	4.0	3.7	–	3.7	3.7
ISR/FSR	2.8	–	–	–	8.0	8.0	8.0	5.2	–	5.2	5.2
b-jet frag.	2.0	–	–	–	2.0	–	–	–	–	2.0	2.0
Jet ID	1.0	–	–	–	1.0	1.0	1.0	1.0	–	1.0	1.0
Jet res.	4.0	–	–	–	4.0	4.0	4.0	4.0	–	4.0	4.0
S_{HF}^{W}	–	13.7	−0.8	13.7	–	–	–	–	–	–	–
S_{HF}^{ratio}	–	−5.0	–	5.0	–	–	–	–	–	–	–
S_{HF}^{Z}	–	–	–	–	13.7	13.7	–	–	–	–	–
IKS	–	1.8	1.8	1.8	–	–	–	–	−30.0	–	–

The shape-shifting systematic uncertainties are not shown in these tables (see Sect. B.2)

Table B.3 Systematic uncertainties for the Run IIa electron channels with four jets

	\multicolumn{11}{l}{Single-tagged four-jet electron channel percentage uncertainties}										
	$t\bar{t}$	$Wb\bar{b}$	$Wc\bar{c}$	Wlp	$Zb\bar{b}$	$Zc\bar{c}$	Zlp	Dibosons	Multijet	tb	tqb
Luminosity	6.1	–	–	–	6.1	6.1	6.1	6.1	–	6.1	6.1
Xsect.	12.7	–	–	–	5.8	5.8	5.8	5.8	–	11.2	7.4
Branching frac.	1.5	–	–	–	–	–	–	–	–	1.5	1.5
PDF	–	–	–	–	–	–	–	–	–	3.0	3.0
Triggers	5.0	–	–	–	5.0	5.0	5.0	5.0	–	5.0	5.0
Lumi. rewtg.	1.0	–	–	–	1.0	1.0	1.0	1.0	–	1.0	1.0
Prim. vertex	1.4	–	–	–	1.4	1.4	1.4	1.4	–	1.4	1.4
Lepton ID	2.5	–	–	–	2.5	2.5	2.5	2.5	–	2.5	2.5
Jet frag.	0.7	–	–	–	4.0	4.0	4.0	4.7	–	4.7	4.7
ISR/FSR	0.6	–	–	–	8.0	8.0	8.0	12.6	–	12.6	12.6
b-jet frag.	2.0	–	–	–	2.0	–	–	–	–	2.0	2.0
Jet ID	1.0	–	–	–	1.0	1.0	1.0	1.0	–	1.0	1.0
Jet res.	4.0	–	–	–	4.0	4.0	4.0	4.0	–	4.0	4.0
S_{HF}^{W}	–	13.7	−0.7	13.7	–	–	–	–	–	–	–
S_{HF}^{ratio}	–	−5.0	–	5.0	–	–	–	–	–	–	–
S_{HF}^{Z}	–	–	–	–	13.7	13.7	–	–	–	–	–
IKS	–	1.8	1.8	1.8	–	–	–	–	−30.0	–	–

	\multicolumn{11}{l}{Double-tagged four-jet electron channel percentage uncertainties}										
	$t\bar{t}$	$Wb\bar{b}$	$Wc\bar{c}$	Wlp	$Zb\bar{b}$	$Zc\bar{c}$	Zlp	Dibosons	Multijet	tb	tqb
Luminosity	6.1	–	–	–	6.1	6.1	6.1	6.1	–	6.1	6.1
Xsect.	12.7	–	–	–	5.8	5.8	5.8	5.8	–	11.2	7.4
Branching frac.	1.5	–	–	–	–	–	–	–	–	1.5	1.5
PDF	–	–	–	–	–	–	–	–	–	3.0	3.0
Triggers	5.0	–	–	–	5.0	5.0	5.0	5.0	–	5.0	5.0
Lumi. rewtg.	1.0	–	–	–	1.0	1.0	1.0	1.0	–	1.0	1.0
Prim. vertex	1.4	–	–	–	1.4	1.4	1.4	1.4	–	1.4	1.4
Lepton ID	2.5	–	–	–	2.5	2.5	2.5	2.5	–	2.5	2.5
Jet frag.	0.7	–	–	–	4.0	4.0	4.0	4.7	–	4.7	4.7
ISR/FSR	0.6	–	–	–	8.0	8.0	8.0	12.6	–	12.6	12.6
b-jet frag.	2.0	–	–	–	2.0	–	–	–	–	2.0	2.0
Jet ID	1.0	–	–	–	1.0	1.0	1.0	1.0	–	1.0	1.0
Jet res.	4.0	–	–	–	4.0	4.0	4.0	4.0	–	4.0	4.0
S_{HF}^{W}	–	13.7	−0.7	13.7	–	–	–	–	–	–	–
S_{HF}^{ratio}	–	−5.0	–	5.0	–	–	–	–	–	–	–
S_{HF}^{Z}	–	–	–	–	13.7	13.7	–	–	–	–	–
IKS	–	1.8	1.8	1.8	–	–	–	–	−30.0	–	–

The shape-shifting systematic uncertainties are not shown in these tables (see Sect. B.2)

Table B.4 Systematic uncertainties for the Run IIa muon channels with two jets

	Single-tagged two-jet muon channel percentage uncertainties										
	$t\bar{t}$	$Wb\bar{b}$	$Wc\bar{c}$	Wlp	$Zb\bar{b}$	$Zc\bar{c}$	Zlp	Dibosons	Multijet	tb	tqb
Luminosity	6.1	–	–	–	6.1	6.1	6.1	6.1	–	6.1	6.1
Xsect.	12.7	–	–	–	5.8	5.8	5.8	5.8	–	11.2	7.4
Branching frac.	1.5	–	–	–	–	–	–	–	–	1.5	1.5
PDF	–	–	–	–	–	–	–	–	–	3.0	3.0
Triggers	5.0	–	–	–	5.0	5.0	5.0	5.0	–	5.0	5.0
Lumi. rewtg.	1.0	–	–	–	1.0	1.0	1.0	1.0	–	1.0	1.0
Prim. vertex	1.4	–	–	–	1.4	1.4	1.4	1.4	–	1.4	1.4
Lepton ID	2.5	–	–	–	2.5	2.5	2.5	2.5	–	2.5	2.5
Jet frag.	0.7	–	–	–	4.0	4.0	4.0	0.7	–	0.7	0.7
ISR/FSR	3.0	–	–	–	8.0	8.0	8.0	0.6	–	0.6	0.6
b-jet frag.	2.0	–	–	–	2.0	–	–	–	–	2.0	2.0
Jet ID	1.0	–	–	–	1.0	1.0	1.0	1.0	–	1.0	1.0
Jet res.	4.0	–	–	–	4.0	4.0	4.0	4.0	–	4.0	4.0
S^W_{HF}	–	13.7	–0.8	13.7	–	–	–	–	–	–	–
S^{ratio}_{HF}	–	–5.0	–	5.0	–	–	–	–	–	–	–
S^Z_{HF}	–	–	–	–	20.0	20.0	–	–	–	–	–
IKS	–	1.8	1.8	1.8	–	–	–	–	–40.0	–	–

	Double-tagged two-jet muon channel percentage uncertainties										
	$t\bar{t}$	$Wb\bar{b}$	$Wc\bar{c}$	Wlp	$Zb\bar{b}$	$Zc\bar{c}$	Zlp	Dibosons	Multijet	tb	tqb
Luminosity	6.1	–	–	–	6.1	6.1	6.1	6.1	–	6.1	6.1
Xsect.	12.7	–	–	–	5.8	5.8	5.8	5.8	–	11.2	7.4
Branching frac.	1.5	–	–	–	–	–	–	–	–	1.5	1.5
PDF	–	–	–	–	–	–	–	–	–	3.0	3.0
Triggers	5.0	–	–	–	5.0	5.0	5.0	5.0	–	5.0	5.0
Lumi. rewtg.	1.0	–	–	–	1.0	1.0	1.0	1.0	–	1.0	1.0
Prim. vertex	1.4	–	–	–	1.4	1.4	1.4	1.4	–	1.4	1.4
Lepton ID	2.5	–	–	–	2.5	2.5	2.5	2.5	–	2.5	2.5
Jet frag.	0.7	–	–	–	4.0	4.0	4.0	0.7	–	0.7	0.7
ISR/FSR	3.0	–	–	–	8.0	8.0	8.0	0.6	–	0.6	0.6
b-jet frag.	2.0	–	–	–	2.0	–	–	–	–	2.0	2.0
Jet ID	1.0	–	–	–	1.0	1.0	1.0	1.0	–	1.0	1.0
Jet res.	4.0	–	–	–	4.0	4.0	4.0	4.0	–	4.0	4.0
S^W_{HF}	–	13.7	–0.8	13.7	–	–	–	–	–	–	–
S^{ratio}_{HF}	–	–5.0	–	5.0	–	–	–	–	–	–	–
S^Z_{HF}	–	–	–	–	20.0	20.0	–	–	–	–	–
IKS	–	1.8	1.8	1.8	–	–	–	–	–40.0	–	–

The shape-shifting systematic uncertainties are not shown in these tables (see Sect. B.2)

Table B.5 Systematic uncertainties for the Run IIa muon channels with three jets

	Single-tagged three-jet muon channel percentage uncertainties										
	$t\bar{t}$	$Wb\bar{b}$	$Wc\bar{c}$	Wlp	$Zb\bar{b}$	$Zc\bar{c}$	Zlp	Dibosons	Multijet	tb	tqb
Luminosity	6.1	–	–	–	6.1	6.1	6.1	6.1	–	6.1	6.1
Xsect.	12.7	–	–	–	5.8	5.8	5.8	5.8	–	11.2	7.4
Branching frac.	1.5	–	–	–	–	–	–	–	–	1.5	1.5
PDF	–	–	–	–	–	–	–	–	–	3.0	3.0
Triggers	5.0	–	–	–	5.0	5.0	5.0	5.0	–	5.0	5.0
Lumi. rewtg.	1.0	–	–	–	1.0	1.0	1.0	1.0	–	1.0	1.0
Prim. vertex	1.4	–	–	–	1.4	1.4	1.4	1.4	–	1.4	1.4
Lepton ID	2.5	–	–	–	2.5	2.5	2.5	2.5	–	2.5	2.5
Jet frag.	0.1	–	–	–	4.0	4.0	4.0	3.7	–	3.7	3.7
ISR/FSR	2.8	–	–	–	8.0	8.0	8.0	5.2	–	5.2	5.2
b-jet frag.	2.0	–	–	–	2.0	–	–	–	–	2.0	2.0
Jet ID	1.0	–	–	–	1.0	1.0	1.0	1.0	–	1.0	1.0
Jet res.	4.0	–	–	–	4.0	4.0	4.0	4.0	–	4.0	4.0
S_{HF}^{W}	–	13.7	−0.8	13.7	–	–	–	–	–	–	–
S_{HF}^{ratio}	–	−5.0	–	5.0	–	–	–	–	–	–	–
S_{HF}^{Z}	–	–	–	–	20.0	20.0	–	–	–	–	–
IKS	–	1.8	1.8	1.8	–	–	–	–	−30.0	–	–

	Double-tagged three-jet muon channel percentage Incertainties										
	$t\bar{t}$	$Wb\bar{b}$	$Wc\bar{c}$	Wlp	$Zb\bar{b}$	$Zc\bar{c}$	Zlp	dibosons	multijet	tb	tqb
Luminosity	6.1	–	–	–	6.1	6.1	6.1	6.1	–	6.1	6.1
Xsect.	12.7	–	–	–	5.8	5.8	5.8	5.8	–	11.2	7.4
Branching frac.	1.5	–	–	–	–	–	–	–	–	1.5	1.5
PDF	–	–	–	–	–	–	–	–	–	3.0	3.0
Triggers	5.0	–	–	–	5.0	5.0	5.0	5.0	–	5.0	5.0
Lumi. rewtg.	1.0	–	–	–	1.0	1.0	1.0	1.0	–	1.0	1.0
Prim. vertex	1.4	–	–	–	1.4	1.4	1.4	1.4	–	1.4	1.4
Lepton ID	2.5	–	–	–	2.5	2.5	2.5	2.5	–	2.5	2.5
Jet frag.	0.1	–	–	–	4.0	4.0	4.0	3.7	–	3.7	3.7
ISR/FSR	2.8	–	–	–	8.0	8.0	8.0	5.2	–	5.2	5.2
b-jet frag.	2.0	–	–	–	2.0	–	–	–	–	2.0	2.0
Jet ID	1.0	–	–	–	1.0	1.0	1.0	1.0	–	1.0	1.0
Jet res.	4.0	–	–	–	4.0	4.0	4.0	4.0	–	4.0	4.0
S_{HF}^{W}	–	13.7	−0.8	13.7	–	–	–	–	–	–	–
S_{HF}^{ratio}	–	−5.0	–	5.0	–	–	–	–	–	–	–
S_{HF}^{Z}	–	–	–	–	20.0	20.0	–	–	–	–	–
IKS	–	1.8	1.8	1.8	–	–	–	–	−30.0	–	–

The shape-shifting systematic uncertainties are not shown in these tables (see Sect. B.2)

Table B.6 Systematic uncertainties for the Run IIa muon channels with four jets

	Single-tagged four-jet muon channel percentage uncertainties										
	$t\bar{t}$	$Wb\bar{b}$	$Wc\bar{c}$	Wlp	$Zb\bar{b}$	$Zc\bar{c}$	Zlp	dibosons	multijet	tb	tqb
Luminosity	6.1	–	–	–	6.1	6.1	6.1	6.1	–	6.1	6.1
Xsect.	12.7	–	–	–	5.8	5.8	5.8	5.8	–	11.2	7.4
Branching frac.	1.5	–	–	–	–	–	–	–	–	1.5	1.5
PDF	–	–	–	–	–	–	–	–	–	3.0	3.0
Triggers	5.0	–	–	–	5.0	5.0	5.0	5.0	–	5.0	5.0
Lumi. rewtg.	1.0	–	–	–	1.0	1.0	1.0	1.0	–	1.0	1.0
Prim. vertex	1.4	–	–	–	1.4	1.4	1.4	1.4	–	1.4	1.4
Lepton ID	2.5	–	–	–	2.5	2.5	2.5	2.5	–	2.5	2.5
Jet frag.	0.7	–	–	–	4.0	4.0	4.0	4.7	–	4.7	4.7
ISR/FSR	0.6	–	–	–	8.0	8.0	8.0	12.6	–	12.6	12.6
b-jet frag.	2.0	–	–	–	2.0	–	–	–	–	2.0	2.0
Jet ID	1.0	–	–	–	1.0	1.0	1.0	1.0	–	1.0	1.0
Jet res.	4.0	–	–	–	4.0	4.0	4.0	4.0	–	4.0	4.0
S_{HF}^{W}	–	13.7	−0.7	13.7	–	–	–	–	–	–	–
S_{HF}^{ratio}	–	−5.0	–	5.0	–	–	–	–	–	–	–
S_{HF}^{Z}	–	–	–	–	20.0	20.0	–	–	–	–	–
IKS	–	1.8	1.8	1.8	–	–	–	–	−30.0	–	–

	Double-tagged four-jet muon channel percentage uncertainties										
	$t\bar{t}$	$Wb\bar{b}$	$Wc\bar{c}$	Wlp	$Zb\bar{b}$	$Zc\bar{c}$	Zlp	Dibosons	Multijet	tb	tqb
Luminosity	6.1	–	–	–	6.1	6.1	6.1	6.1	–	6.1	6.1
Xsect.	12.7	–	–	–	5.8	5.8	5.8	5.8	–	11.2	7.4
Branching frac.	1.5	–	–	–	–	–	–	–	–	1.5	1.5
PDF	–	–	–	–	–	–	–	–	–	3.0	3.0
Triggers	5.0	–	–	–	5.0	5.0	5.0	5.0	–	5.0	5.0
Lumi. rewtg.	1.0	–	–	–	1.0	1.0	1.0	1.0	–	1.0	1.0
Prim. vertex	1.4	–	–	–	1.4	1.4	1.4	1.4	–	1.4	1.4
Lepton ID	2.5	–	–	–	2.5	2.5	2.5	2.5	–	2.5	2.5
Jet frag.	0.7	–	–	–	4.0	4.0	4.0	4.7	–	4.7	4.7
ISR/FSR	0.6	–	–	–	8.0	8.0	8.0	12.6	–	12.6	12.6
b-jet frag.	2.0	–	–	–	2.0	–	–	–	–	2.0	2.0
Jet ID	1.0	–	–	–	1.0	1.0	1.0	1.0	–	1.0	1.0
Jet res.	4.0	–	–	–	4.0	4.0	4.0	4.0	–	4.0	4.0
S_{HF}^{W}	–	13.7	−0.7	13.7	–	–	–	–	–	–	–
S_{HF}^{ratio}	–	−5.0	–	5.0	–	–	–	–	–	–	–
S_{HF}^{Z}	–	–	–	–	20.0	20.0	–	–	–	–	–
IKS	–	1.8	1.8	1.8	–	–	–	–	−30.0	–	–

The shape-shifting systematic uncertainties are not shown in these tables (see Sect. B.2)

B.2 Shape-Changing Systematics

To evaluate the uncertainties on the jet energy scale and the b-tagging efficiency, four additional sets of simulated samples are produced with these quantities shifted up and down by one standard deviation of their uncertainty. The W+jets samples are also re-reproduced with the all ALPGEN reweightings (Sect. 5.4.4) shifted up and down by one standard deviation. The yield subsets (Sect. 7.1.1) of these six samples are passed through the final boosted decision trees, and new discriminant output histograms are produced. Some of these histograms from the 1tag-2jet channels are seen in Figs. B.1 and B.2, and from the 2tag-2jet channels in in Figs. B.1 and B.2. The difference between the histogram is the uncertainty used in the cross-section calculation as explained in Sect. 7.2.2.

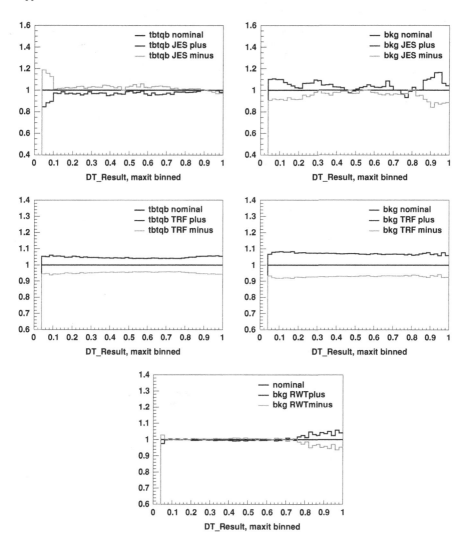

Fig. B.1 Shape-shifting systematics for Run IIb, e+jets, 2jets-1tag. The boosted decision tree distributions are produced from the nominal and shifted samples. We have single *top* (*left*) and all backgrounds combined (*right*) for jet energy scale (*top row*), b tagging (*middle row*) and ALPGEN reweighing for W+jets only (*bottom plot*)

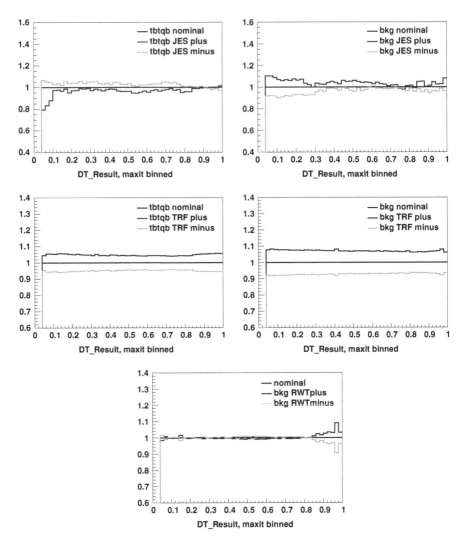

Fig. B.2 Shape-shifting systematics for Run IIb, μ+jets, 2jets-1tag channel. The boosted decision tree distributions are produced from the nominal and shifted samples. We have single top (*left*) and all backgrounds combined (*right*) for jet energy scale (*top row*), *b* tagging (*middle row*) and ALPGEN reweighing for *W*+jets only (*bottom plot*)

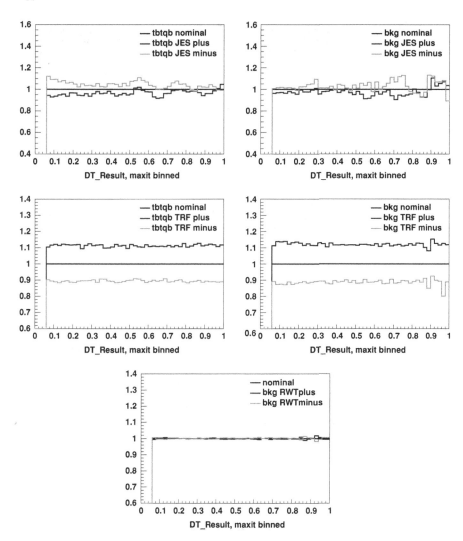

Fig. B.3 Shape-shifting systematics for Run IIb, *e*+jets, 2jets-2tag channel. The boosted decision tree distributions are produced from the nominal and shifted samples. We have single top (*left*) and all backgrounds combined (*right*) for jet energy scale (*top row*), *b* tagging (*middle row*) and ALPGEN reweighing for *W*+jets only (*bottom plot*)

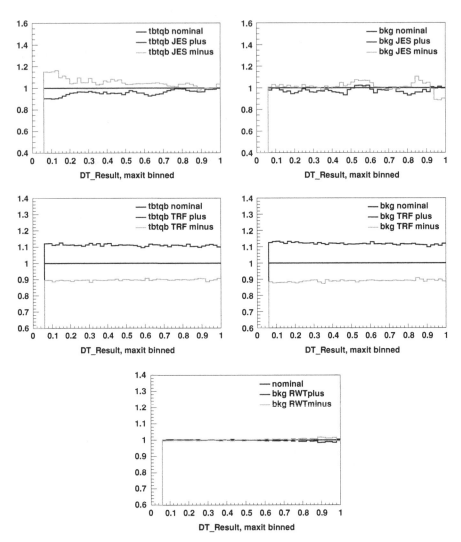

Fig. B.4 Shape-shifting systematics for Run IIb, μ+jets, 2jets-2tag channel. The boosted decision tree distributions are produced from the nominal and shifted samples. We have single top (*left*) and all backgrounds combined (*right*) for jet energy scale (*top row*), b tagging (*middle row*) and ALPGEN reweighing for W+jets only (*bottom plot*)

Appendix C

C.1 Decision Tree Outputs

This appendix presents the boosted decision tree output distributions for all of the 24 individual channels. Each distribution is shown both using linear and log scale of the y-axis.

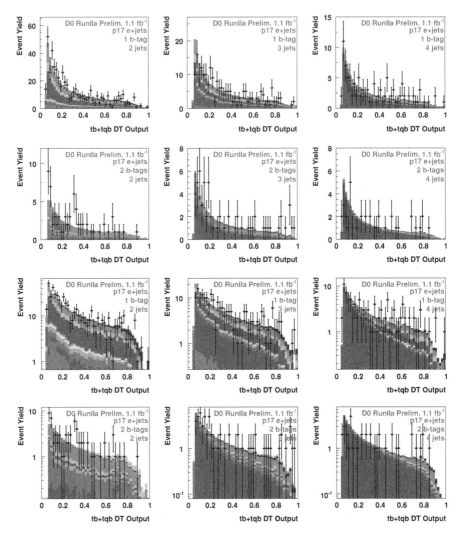

Fig. C.1 Boosted decision tree discriminant output distributions for the six Run IIa *e*+jets channels with two (*left column*), three (*middle column*) and four (*right column*) jets and one and two *b*-tagged jets (*alternating rows*) using linear scale (*top two rows*) and log scale (*bottom two rows*). The plot key can be seen in Fig. 5.6

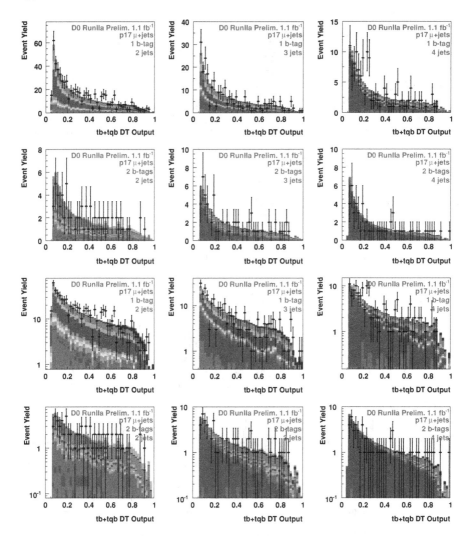

Fig. C.2 Boosted decision tree discriminant output distributions for the six Run IIa μ+jets channels with two (*left column*), three (*middle column*) and four (*right column*) jets and one and two b-tagged jets (*alternating rows*) using linear scale (*top two rows*) and log scale (*bottom two rows*). The plot key can be seen in Fig. 5.6

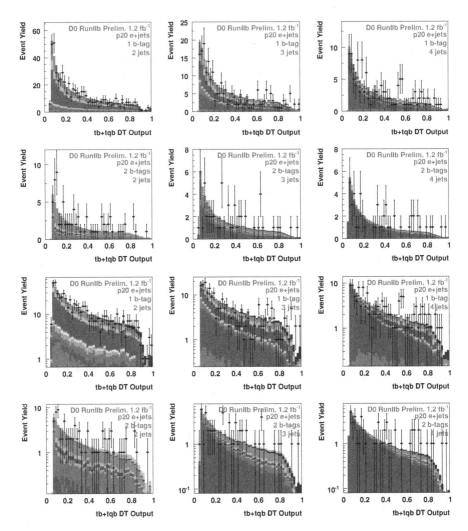

Fig. C.3 Boosted decision tree discriminant output distributions for the six Run IIb μ+jets channels with two (*left column*), three (*middle column*) and four (*right column*) jets and one and two b-tagged jets (*alternating rows*) using linear scale (*top two rows*) and log scale (*bottom two rows*). The plot key can be seen in Fig. 5.6

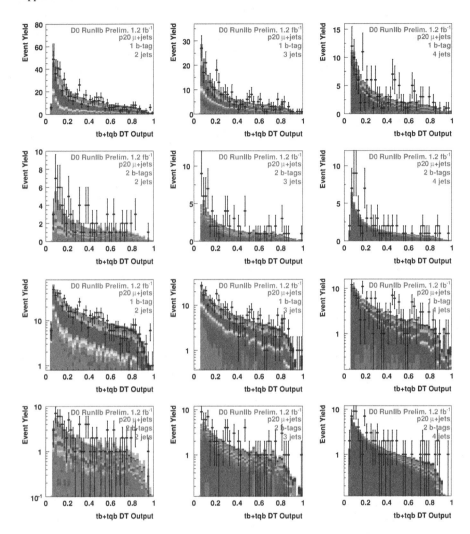

Fig. C.4 Boosted decision tree discriminant output distributions for the six Run IIb μ+jets channels with two (*left column*), three (*middle column*) and four (*right column*) jets and one and two b-tagged jets (*alternating rows*) using linear scale (*top two rows*) and log scale (*bottom two rows*). The plot key can be seen in Fig. 5.6

Appendix D

D.1 Cross Check Samples

This Appendix presents the boosted decision tree output distributions for two W+jets and $t\bar{t}$ cross check samples in Figs. D.1 and D.2, respectively. Figure D.3 shows the boosted decision tree output for the data, signal and background samples before any b-tagging selection is applied.

The W+jets and $t\bar{t}$ cross check samples are defined as follows:

- "W+jets" (2 jets, 1 tag, $H_T < 175$ GeV),
- "$t\bar{t}$" (4 jets, 1–2 tags, $H_T > 300$ GeV).

Fig. D.1 Boosted decision tree discriminant output distributions for the "W+jets" sample for the e+jets (*left*) and μ+jets (*right*) and Run IIa (*top row*) and Run IIb (*bottom row*) channels. The plot key can be seen in Fig. 5.6

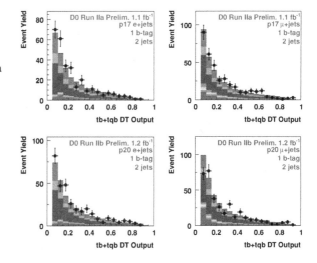

Fig. D.2 Boosted decision tree discriminant output distributions for the $t\bar{t}$ sample for the e+jets (*left*) and μ+jets (*right*) and Run IIa (*top row*) and Run IIb (*bottom row*) channels. The plot key can be seen in Fig. 5.6

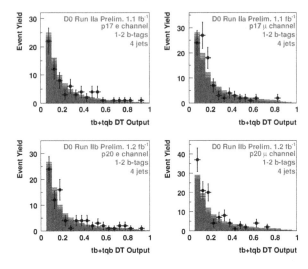

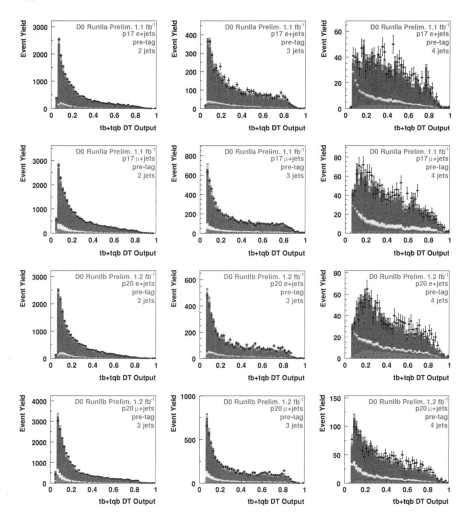

Fig. D.3 Boosted decision tree output distributions for the each of the 12 pre-tag channels, meaning the samples split by run period, jet and tag multiplicity before any *b*-tagging selection is applied. p17 and p20 refers to the Run IIa and Run IIb run periods, respectively. The plot key can be seen in Fig. 5.6

Appendix E

E.1 Combined Results

This appendix presents the results from the other multivariate analyses that were conducted using the same data and simulated samples as the boosted decision tree analysis presented in this thesis.,

The other two individual analyses use Bayesian neural networks (BNN) [1, 2] and the Matrix Element method (ME) [3, 4] to separate single top from backgrounds. Both of these analyses are updated versions of the previous analyses [5], which established the first evidence for single top quark production in 2006.

As discussed in Sect. 7.4.1, the three individual multivariate outputs were used as input to a second layer of Bayesian neural network. The resulting super discriminant (BNN combination output) is more powerful than any of the discriminants for the individual analyses. The discriminant outputs for the individual

Fig. E.1 The discriminant output distribution for all channels combined for boosted decision trees (**a**), Bayesian neural networks (**b**), the matrix element method (**c**) and the BNN combination (**d**). The BDT distribution is the same shown in Fig. 7.19, but a different binning is used

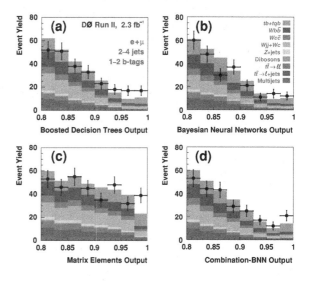

Table E.1 Expected and observed significances and measured single top cross sections for the three different multivariate techniques and their combination

Multivariate analysis results

Analysis	Significance		Measured
	Expected	Observed	σ_{s+t} [pb]
Boosted decision tree	**4.3σ**	4.6σ	**3.7**$^{+1.0}_{-0.8}$
Bayesian neural networks	4.1σ	5.2σ	4.7$^{+0.9}_{-0.9}$
Matrix elements	4.1σ	5.0σ	4.3$^{+0.9}_{-0.9}$
BNN combination	**4.5σ**	**5.0σ**	**3.9**$^{+0.9}_{-0.9}$

The boosted decision tree analysis is the most sensitive of the ordinary multivariate analyses with an expected significance of 4.3σ. This significance improves as the analyses are combined. The observed significance for the BNN combination is 5σ, corresponding to the first observation of single top quark production

multivariate techniques, as well as the combination, are presented in Fig. E.1, and the cross section and significance measurements are shown in Table E.1. The boosted decision tree analysis is more sensitive than the BNN and ME analyses, but not as sensitive as the BNN combination.

The BNN combination output is also used to derive a cross section measurement yielding: $|f_1^L V_{tb}| = 1.07 \pm 0.12$, and $|V_{tb}| > 0.78$ at the 95% confidence level.

References

1. R.M. Neal, *Bayesian Learning for Neural Networks* (Springer, New York, 1996)
2. A. Tanasijczuk, Ph.D. thesis, Single top quark cross section measurement in proton–antiproton collisions at $\sqrt{s} = 1.96$. (Universidad de Buenos Aires, in preparation)
3. V.M. Abazov et al. (DØ Collaboration), A precision measurement of the mass of the top quark. Nature **429**, 638 (2004)
4. M. Pangilinan, Ph.D. thesis, Observation of single top quark production at DØ using matrix elements. (Brown University, in preparation)
5. V.M. Abazov, et al. (DØ Collaboration), Evidence for production of single top quarks. Phys. Rev. D **78**, 012005 (2008)